Core Teachings of the Dalai Lama

Where Buddhism Meets Neuroscience

Conversations with the Dalai Lama on the Spiritual and Scientific Views of Our Minds

EDITED BY

Zara Houshmand, Robert B. Livingston, and B. Alan Wallace

WITH CONTRIBUTIONS BY

Patricia Smith Churchland, PhD; Antonio R. Damasio, MD; J. Allan Hobson, MD; Lewis L. Judd, MD; and Larry R. Squire, PhD

TRANSLATED BY

Geshe Thupten Jinpa and B. Alan Wallace

SHAMBHALA
BOULDER
2018

Shambhala Publications, Inc.
4720 Walnut Street
Boulder, Colorado 80301
www.shambhala.com

The book was previously published under the title *Consciousness at the Crossroads*.

9 8 7 6 5 4 3 2 1

Printed in the United States of America

♾ This edition is printed on acid-free paper that meets the American National Standards Institute z39.48 Standard.
♻ This book is printed on 30% postconsumer recycled paper. For more information please visit www.shambhala.com.

Shambhala Publications is distributed worldwide by Penguin Random House, Inc., and its subsidiaries.

Library of Congress Cataloging-in-Publication Data

Names: Bstan-'dzin-rgya-mtsho, Dalai Lama XIV, 1935– author. | Houshmand, Zara, editor. | Livingston, Robert B. (Robert Burr), 1918–2002, editor. | Wallace, B. Alan, editor, translator. | Thupten Jinpa, translator. | Churchland, Patricia Smith. | Damasio, Antonio R. | Hobson, J. Allan, 1933– | Judd, Lewis L. | Squire, Larry R. | Mind & Life Institute.

Title: Where Buddhism meets neuroscience: conversations with the Dalai Lama on the spiritual and scientific views of our minds/edited by Zara Houshmand, Robert B. Livingston, and B. Alan Wallace; with contributions by Patricia Smith Churchland, PhD; Antonio R. Damasio, MD; J. Allan Hobson, MD; Lewis L. Judd, MD; and Larry R. Squire, PhD; translated by Geshe Thupten Jinpa and B. Alan Wallace.

Other titles: Consciousness at the crossroads

Description: Boulder: Shambhala, 2018. | Series: Core teachings of the Dalai Lama | Includes bibliographical references and index. | "The book was previously published under the title Consciousness at the Crossroads."

Identifiers: LCCN 2018007116 | ISBN 9781559394789 (paperback)

Subjects: LCSH: Consciousness—Religious aspects—Buddhism. | Consciousness—Physiological aspects. | Brain—Psychophysiology. | Buddhism—Psychology. | BISAC: RELIGION/Buddhism/Tibetan. | PHILOSOPHY/Buddhist. | SCIENCE/Life Sciences/Neuroscience.

Classification: LCC BQ7935.B774 W49 2018 | DDC 294.3/3615—dc23

LC record available at https://lccn.loc.gov/2018007116

Contents

Introduction

On the morning of October 5, 1989, history intruded unexpectedly into a private meeting between His Holiness the 14th Dalai Lama of Tibet and a small group of neuroscientists and psychiatrists. Leaders in their fields, they had come together to explore what insights the Western sciences of the mind and Buddhism might offer to each other. The second Mind and Life Conference was gathered at the Newport Beach home of Mr. and Mrs. Clifford Heinz, when a predawn phone call from Oslo announced that His Holiness had been awarded the Nobel Prize for Peace.

Considering how public awareness of the situation in Tibet has grown in the West in recent years, it is easy to forget how significant the Dalai Lama's Nobel Prize was at the time. After decades of international neglect, the award was an important turning point for the Tibetan cause, recognizing the Tibetan people's long struggle against Chinese oppression as well as the Dalai Lama's commitment to a nonviolent resolution of the continuing conflict.

Shortly after the first phone call, other calls began coming in from the television networks. By seven o'clock, His Holiness had made the decision to continue with the conference as scheduled, and two hours later the group convened. By the time His Holiness entered the living room, which had been rearranged for the conference, and took his seat in the circle, an extraordinary sense of joy and excitement had filled the house. Robert Livingston, the scientific coordinator, spoke a few warm words of congratulation. His Holiness responded to the effect that the prize should not be considered as a recognition of any personal qualities of his own, but was important as a recognition of the path of nonviolence he followed.

Even those closest to him, who well knew the Dalai Lama's characteristic humility, were surprised at his nonchalance that day—and the more so for their own excitement. Those who were meeting him that day for the first time were profoundly struck by his equanimity at receiving this highest honor of humanity.

The Dalai Lama's decision to proceed with the conference as planned that day is evidence of the importance these dialogues hold for him. The first Mind and Life Conference had met two years earlier, in October 1987.[1] The meetings were initiated jointly by Adam Engle, a U.S. attorney and businessman, and Dr. Francisco Varela, a neurobiologist at the National Center for Scientific Research, in Paris, in response to His Holiness's lifelong interest in the sciences, and a growing awareness of the potential for a serious dialogue between Buddhism and Western science.

The conferences would meet every two years, usually in the remote but idyllic setting of the Dalai Lama's home in Dharamsala, India, for five full days each time. The first conference had provided a broad overview of the mind sciences, with presentations on scientific method, perception, cognitive psychology, artificial intelligence, developmental neurobiology, and evolution. In years to come, the third and fourth conferences[2] would continue the emphasis on mind sciences, focusing first on the effect of emotions on health and then on sleep, dreaming, and dying. The fifth conference in 1995 moved further afield, into the study of compassion, altruism, and ethics. Most recently, in 1997, the dialogue has moved in a new direction, addressing the new physics, cosmology, and quantum mechanics.

The second conference, reported here, was unusual among the series in that it was only two days long and took place in the West, in Newport Beach, California. Dr. Robert Livingston, MD, Professor Emeritus of Neurosciences at the University of California, San Diego, who had been invited to participate in the first Mind and Life dialogue two years before, took on the responsibility of being the scientific coordinator for this conference. The colleagues he selected to represent their fields were outstanding individually and formidable as a group.

Patricia Smith Churchland, PhD, professor of philosophy at the University of California at San Diego, set the context of the dialogue in the philosophical and historical origins of Western sciences of the mind. Antonio R. Damasio, MD, professor of neurology at the University of Iowa College of Medicine, reviewed findings on the relationship between the anatomy of the brain and mental functions. Larry R. Squire, PhD, professor of psychiatry at the University of California at San Diego, introduced the science of memory. J. Allan Hobson, MD, professor of psychiatry at Harvard Medical School, provided an overview of current knowledge on sleep and dream states. And Lewis L. Judd, MD, then director of the National Institute of Mental Health, outlined current views on mental illness and psychopharmacology. Translation was provided by Thupten Jinpa and B. Alan Wallace.

The format of the Mind and Life Conferences consists of formal presentations from each of the participating scientists and philosophers, alternating with open-ended discussion. In the context of this dialogue, the scientists are committed to representing consensus in their respective fields, as this is not seen as an appropriate forum for airing controversial material or furthering debate within the academy.

The presentations are interrupted often, as the Dalai Lama asks a question or offers an immediate response to a point. The discussion is often sparked by the preceding presentation, but may draw on topics from previous conversations. In the account that follows, portions of dialogue from different sessions may be grouped together, tracing themes that developed gradually over the course of the meeting.

Throughout the meetings, His Holiness listens intently to each speaker, following most of the English, though occasionally turning to the translators to ask for clarification. In his own responses, he usually speaks through the translators when dealing with Buddhist philosophy or scientific concepts. But he often breaks into English to communicate ideas that are less technical—to express his feelings, to make a joke, or to describe his own experiences—and these are moments of special warmth.

In preparing the text of this book, the contributions of the translators, Thupten Jinpa and Alan Wallace, have been rendered transparent, except on rare occasions as they voice their own concerns about the communication in process. Thus, when His Holiness speaks through a translator, the speech is represented as his own.

In addition to his role as a translator, Alan Wallace has contributed significantly to the shape of this book by adding commentary to clarify the Buddhist viewpoint on issues that were raised at the meeting but not well elucidated at the time. We hope that the value of this commentary, especially as it touches on points that are easily misconstrued in the cross-cultural dialogue, will outweigh the dangers of allowing one side of the debate to add a last word after the meeting has ended. But of course, the dialogue between Buddhism and Western science continues, and this book is only a snapshot of one particularly lively moment in the discourse.

Zara Houshmand

Where Buddhism Meets Neuroscience

1. Opening Remarks

Brain Science as a Path to World Peace

Robert Livingston opened the morning session on the first day of the conference by voicing the general elation at the news, just announced, that the Dalai Lama had been awarded the Nobel Peace Prize. In his role as the scientific coordinator, Dr. Livingston articulated his personal vision of the purpose of the dialogue: that a better understanding of the brain's complexity and adaptability, and of the resulting diversity of human consciousness, is critical to global human survival In the light of the Nobel Prize announcement, his message carried a profound sense of the responsibility of each of the participants gathered in that room. This face-to-face engagement of Western science with the Buddhist tradition of enquiry into consciousness might well have deep ramifications for world peace.

ROBERT LIVINGSTON: This initiates the second dialogue between Western neurosciences and Buddhist traditions. These two radically different ways of looking at mind and life have existed, mostly apart, over a span of about 2,500 years. They have been following such separate paths that there has been almost no cross-communication. So for all of us this is a significant opportunity. We anticipate that the Mind and Life dialogues will improve and increase communications and strengthen ties in terms of mutual understanding of neurosciences, consciousness, brain, mind, and the like, and also add new insights into human nature which we believe can contribute to world peace.

There are indeed two great fundamentals underlying this dialogue. First, the issues up for discussion here are not only of great importance in each individual human life, but their comprehension by a wider public may indeed be pivotal for human survival on a global scale. Such issues relate to individual and collective differences in perception, judgment, behavior, and communication. And second, the human brain is the only resourceful instrument for survival. It has always been obliged to be—and continues to be—constructively adaptive. Yet its full potential will not be realized until the brain is better understood, particularly in terms of its individuality and the consequent diversity of worldviews.

The brain is constructively adaptive in the sense that it is continually self-actuating and self-organizing with respect to its own body, and its projection and testing of images and maps of the outside world. According to its own timetable, it changes its internal states and partitions its activities swiftly and in a comprehensively integrated way.

There is a prior, slower brain dynamic in evolution, shaped by selective forces acting over extremely long periods of time. In all history, the most abrupt response to selection forces affecting brain evolution was the approximately threefold volumetric expansion of the hominid brain, which began about three million years ago with a common ancestor from which we diverged from present-day chimpanzees. In periods of individual human lifetimes, the brain is dynamic, too, in its embryonic, fetal, and childhood development, including its remarkable adaptation to a given environment and culture, and its diminishing powers associated with disease and aging.

These evolutionary and life-span changes are structurally dynamic at gross, microscopic, and ultrastructural levels of neuroanatomy. Changes in ultrastructure—at the level of electron microscopy—are occurring dynamically even as we think, talk, behave, and remember events. Changes at microscopic levels of organization take place at a slower rate in accordance with our use or desuetude of particular aspects of our conscious and unconscious experiences. Brain states arise from neuronal activity that involves dynamic bioelectrical and biochemical events, and that can change the ultrastructural features of the fine membranous architectures of cellular neighborhoods.

Very importantly, the brain is dynamic in an integrative sense. Whenever we examine someone with a drugged, diseased, or damaged brain, we observe what that particular brain can do as a whole, despite whatever enduring damage may have occurred, and despite whatever transient interferences may be occurring. In other words, the brain as a whole tends to do the best it can by integrating all of its resources that are available.

The brain is likewise dynamic in a personal, subjective, interactive sense, which I should like to illustrate for you now. For this purpose, I invite your participation so that your Holiness can focus consciously on some subjective experiences as they take place within your own brain.

Tibetan people have undoubtedly gazed at waterfalls often and for prolonged periods. When someone looks at a waterfall, steadily at one point in the falls, for at least a few minutes, and then looks away at the wall of the mountain adjacent to the waterfall, something amazing happens perceptually. Specifically, the wall of rock, in a width corresponding to the waterfall, now appears to move upwards and does so for some minutes.

This tells us that something dynamic has happened within particular brain circuits that process visual perception. Some among them have become temporarily actively engaged in adapting their powers of discriminative analysis to the problems of better perceiving swiftly falling water. We infer that after one has gazed at the waterfall for a short while, the perceptual apparatus has adapted itself so as to slow down the motion of the falling water, perhaps to allow it to be more precisely observed. The evidence is that the slowing down process, which is confined to a well-defined patch of the visual field centered on the waterfall, persists in dynamic fashion after one looks away and operates to produce the reverse motion perception of objects perceived by that transiently adapted sector of the visual field.

You have probably noticed also on a sea voyage, that after much rolling and pitching motion of the ship, a similar sense of motion can persist for many hours after you go ashore. Or following flight in an airplane, your hearing may be affected for some hours after landing. There are many other commonplace means of witnessing dynamic features of brain mechanisms. For example, when you have traveled across several time zones in an airplane, you have undoubtedly experienced jet lag. Your body, including your endocrine system, digestion, and sleep mechanisms, takes some days to recover normal rhythms while your brain is readjusting your daylight cycle entrainment in accordance with the new time zone.

At this point, Dr. Livingston asked His Holiness to participate in an exercise to become consciously aware of one's own brain during voluntary performance. He demonstrated by spreading and closing his fingers in a fanlike motion, and asked the Dalai Lama to copy this action.

When you do this, you are first of all politely acceding to my request that you do so. Then you are engaging in voluntary, deliberately willed actions. This involves certain parts of your motor cortex in a discrete and complicated pattern of electrochemical activity. You can subjectively recognize the feelings generated by your intentions, your initiatives, and the corresponding perceptual experiences that return from your fingers, hands, and vision, to inform you that you have performed the actions more or less appropriately.

Dr. Livingston then took the Dalai Lama's hand in his own, holding his fingers together.

If I restrain your fingers so that your voluntary actions are gently prevented, and ask you to nevertheless continue spreading your fingers in the same way, you can immediately appreciate the difference required by having to make extra brain efforts, or willpower, as well as extra muscular efforts.

DALAI LAMA: So what is the meaning of this?

ROBERT LIVINGSTON: Using Western neuroscientific techniques we could obtain simultaneous objective and subjective evidence that many specific parts of your brain were performing certain orderly, yet very complicated, operations: responding to my request, deciding whether to comply, executing particular finger performances, doing so against resistance, and so forth. Altogether a fantastically large number of distinctive parts of your brain were involved in these activities—auditory, visual, visceromotor, somatomotor, somesthetic, etc.—and individual body parts have been commanded discretely, differentially, and with exquisite sequential precision. In short, for this simple performance, there has been a spectacular orchestration of activities among many discrete parts of your brain. I venture to say that at least a few dozen billion nerve cells and numerous scores of trillions of electrochemical signals were activated in the course of this modest exercise. These brain events involved near and remote circuits and multiple sectors of cortical and subcortical cellular constellations, all of which were harmoniously integrated.

We take this entirely for granted, but it is nonetheless quite astonishing to consider. I suggest that we need to engage in this kind of thinking in order to gain an inkling of the complexities and finesse of human brain processes, and to hold the brain's potentialities in sufficient awe. These are the kinds of considerations that bring Western neuroscientists to determine experimentally how detailed brain processes occur during perception, judgment, and behavior. How are such detailed brain events channeled to control our bodies, to produce sentences, to generate ideas, experience, and manifest emotions? What is the nature of central brain command, and, more broadly, how do dynamic changes of brain state take place between sleep and wakefulness, or between sleep, deep sleep, and dreaming sleep? How are these changes of brain state controlled? More broadly still, what is the role of consciousness in such activities?

These are the kinds of questions we would like to open for free discussion with you in this dialogue. We pay our respects to two cultural traditions which have been separated for so very long and now have a cordial opportunity for exchange by virtue of your curiosity, initiative, and generosity. This opens for neuroscientists an excellent opportunity for professional enrichment because Buddhists have been thinking about consciousness, mind, and body for a very long time along different conceptual paths. We must acknowledge our own humility and naiveté: there are many things that we don't yet know about the brain and the mind, so many about which we are unsure, and others still about which we remain unknowingly in error.

We shall attempt to represent Western neurosciences in a fair way. We can then become your allies in helping insofar as possible to increase mutual understanding in both directions and to dissolve barriers that have too long separated these two insight-seeking cultures. In the process, it should be possible to devise innovative experimental strategies directed to objectifying phenomena studied according to both traditions.

Many fundamental concepts are swiftly changing in Western neurosciences' views relating to brain mechanisms. So we must stay tuned in order to move together within this tumbling stream of scientific innovation.

One of the fundamentals underlying these dialogues is our mutual concern for world peace. We sincerely believe that several pertinent disciplines from both traditions are of great importance for humanity to help in the development of more rational human self-knowledge, cross-cultural mutual understanding, and compassion, all urgently needed to safeguard this planetary habitat and to ensure equitable sharing of its bounty.

2. Toward a Natural Science of the Mind

Patricia Smith Churchland, PhD

As a general rule, each Mind and Life Conference opens with a presentation, often from a philosopher or historian of science, that maps out some of the deep assumptions and cultural givens of the territory to be covered in the dialogue. These buried foundations of our intellectual constructs determine what we allow as legitimately belonging to the realm of science. This is, of course, vital information for anyone coming from a different tradition who seeks to understand Western science on its own terms. This is also the ground we defend irrationally as well as rationally, and where we need to work hardest to open our minds if we are to engage seriously with a very different worldview.

Patricia Churchland's presentation explores the roots of modern Western understanding of the relationship between the mind and the brain. She traces the history of modern science back to its origins in Greek philosophy. The investigations of philosophers such as Aristotle and Plato into the nature of the universe, and of human perception and conceptual representation of that universe, created templates that still have a powerful influence on our own thinking. She describes the dualist position as formulated by Descartes: the separation of body and mind, of physical and spiritual worlds, that has dominated Western thinking for centuries and remains current both in Western religions and in the popular imagination. In opposition to this, she defends the materialist position that underlies most modern scientific thought. In this view, nothing exists other than the physical, and consciousness is understood as an emergent property of the physical organization of the brain.

PATRICIA CHURCHLAND: I think most scientists who work on the brain probably do feel, first, that consciousness is not independent of the brain; second, that memories are a function of the way the brain organizes itself; and third, that perception is dependent upon the way parts of the brain are organized and interact. Part of what I want to try to do is to ask the larger question about the relation of brain and mind.

DALAI LAMA: What do you mean by *consciousness*?

PATRICIA CHURCHLAND: I mean that we are aware of colors, smells, shapes, sounds, and feelings like anger or embarrassment. I don't mean that consciousness must be a *thing*. That's a question to be asked.

I wanted to sketch out how it came to be that many neuroscientists, and I, too, have become convinced that the mind is not independent of the brain—that mind is, in fact, the brain understood as organized and functioning in a certain way.

The Philosophical Roots of Science

I want to start with the ancient Greeks who formulated many concepts and ideas that have been embedded in Western science. There are many scientific questions that originated in the Greek tradition in Western philosophy. Aristotle and Plato were profoundly puzzled about the nature of space and time, and about the nature of substance. They wanted to know what matter is made of, how it is put together. The nature of how things change, and why they change, was also questioned. They noticed that certain things change systematically and other things change apparently at random. They wanted to know about the nature of motion.

Additionally, of course, they wanted to know about the origins of life, where it all came from. They also asked particular questions about humans. They wanted to know how it is that humans can know the world outside them according to some kind of internal representation. How is it that with just a head and eyes and ears, it is nonetheless possible to know what's going on outside in an immense and complicated world, to know that things exist independently of you in the world, and to know that sometimes how things seem to be is different from how they in fact are. These are basic mysteries.

They would use the example of your perception when you thrust a stick into water: the stick looks bent. That is how you perceive it to be, but some other part of your knowledge represents it as being, in fact, straight. They were intrigued by the fact that you could distinguish between how things *seem* to be and how they *actually* turn out to be. They were intrigued, also, that we can think about things when they are not present. Thus, I can think about my children, even though they are not here.

DALAI LAMA: This is a function of the conceptual mind, to create the imagery corresponding to some external event.

PATRICIA CHURCHLAND: But how this is done by the mind/brain is the question. The Greeks noticed that perceptual experience has a kind of unity. They realized that we see shapes and colors as part of one object, such as a tree. There is also unity across different sense modalities. One can see and hear, and realize that both senses arise from the same object. One can see the lyre being played and hear the sound of it, and know that it is part of one thing.

There is also a unity through time. Thus, I know that I am the same person today that I was yesterday and that I was the day before. I can look at this room, scanning different areas of space at different times and yet it remains a unified room for me. It is not a bit here, a bit there, a bit elsewhere as my eyes move. It is understood by me as belonging together, as a whole.

On Perception, Representation, and Conceptualization

DALAI LAMA: When you say *represent*, are you referring only to conceptual impressions or to sensory impressions such as visual and auditory perceptions—or do you mean both?

PATRICIA CHURCHLAND: I mean to cover both. Representations can be perceptual or conceptual. You can have sensory representation from touch, as well as from vision, hearing, and other sense modalities. I would also include memory representations that occur in the course of just thinking about things. We consider all of these brain processes as constituting representations.

When you think about what you are going to say next, and you frame it or mull it over in your head, then you are engaging a linguistic representation. But the critical point is that all of this is occurring in the brain.

DALAI LAMA: Is this function of representation common to all higher animals, at least the more intelligent ones, or is it unique to human beings?

PATRICIA CHURCHLAND: Nobody really knows. But I would suspect that it is found in mammals, birds, and probably also in reptiles and fish.

DALAI LAMA: Do you believe that this is so for both sensory perception and conceptual faculties?

PATRICIA CHURCHLAND: Yes. But for very simple animals, and for us too, there are some very simple behaviors like reflexes. If touching an animal provides an input signal to its nervous system, and this is all that is required to generate a stereotyped response, no mediation beyond some relatively simple reflex pathway is necessary. In a sense, a reflex reflects a very primitive representation. Or you might say, in a very simple case, that there is little or no representation.

DALAI LAMA: Is there not even a sensory representation in that straight reflex situation?

PATRICIA CHURCHLAND: I don't have a decisive answer to that. It is probably not useful to think of it as involving sensory representation. It can take place in the absence of consciousness.

ALLAN HOBSON: It is doubtful whether animals have conceptual representation. It is hard to imagine without language how a concept can arise. I think most scientists would have grave doubts about that.

ROBERT LIVINGSTON: Certainly animals have memories; they have a keen sense of space and time. They have a sense of loyalty.

ALLAN HOBSON: But are those concepts?

PATRICIA CHURCHLAND: It is hard to say in animals. There are of course anecdotes involving chimpanzees who appear to be looking for things that they expect to find in a certain place where they have never found them before. That would indicate that they have a representation.

DALAI LAMA: They do have memory. Isn't memory included in the conceptual faculties? Isn't memory necessarily a conceptual process?

PATRICIA CHURCHLAND: There may be different kinds of memory. Some are conceptual and some, perhaps through habituation, are not. But let me say that the notion of representation is not precise.

ANTONIO DAMASIO: Perhaps it would be useful to think of representation as having many possible levels. There are representations that can correspond to very objective things in the outside world, such as my representation of Pat or of this table. There are other representations that do not relate to the objective world as we perceive it, but are still representations in the sense that they exist in the brain in the form of patterns of activity.

Thus, if you have a reflex, there is in fact a representation of some sort, which corresponds to the sequence of activities that neurons and their synapses perform in order for the reflex to hold a pattern, and to respond characteristically, repeatedly. It is just that it is not a representation in the sense of a picture, an image. But I think we can talk about it as a representation.

And on the issue of conceptualization in animals, I would say that although language adds "quality" to concepts, that animals, certainly animals such as a chimpanzee, do have concepts.

PATRICIA CHURCHLAND: Some of the questions that were of interest to the ancients had to do with the nature of space and time and substance, and so forth. As people began to construct hypotheses about these matters, and as they began to test them experimentally, the questions began to shift into a different domain, to become part of what was established as very probably true, what we call "reality." Thus, very good theories developed about astronomy. It began to be known, certainly by the fifteenth century, that the earth was not in fact the center of the universe, though that disappointed many people. Galileo introduced theories of mechanics; through the work of Newton the nature of motion became well understood; and this was followed in another century by an understanding of heat as molecular motion.

Questions about the nature of substance, what makes up stuff, and how one kind of stuff is different from another shifted over into the domain of chemistry, which likewise became a science, and so also with biology. Issues that used to belong to natural philosophy became part of natural science and were investigated experimentally; hypotheses were tested and technology advanced.

Just as some of the other great questions became scientific as hypotheses and technology progressed, can we also develop a natural science of the mind? This has now become an important contemporary question.

Some scientific work has been done at the level of psychology. For example, we try to characterize a capacity, such as the capacity for seeing in depth, in terms of behavioral responses. But of course what we really want to know is, how is that possible? What is the mechanism for seeing in depth? We now know how accurate depth perception can be; we know some of the scope and limits of that capacity, but what we seek to understand is: How does it work? How can something produce depth perception?

And similarly with memory: How is it possible that we can be exposed to something, and then remember it a day later, or a year later, or ten years later? How does memory work?

Mind and Brain: One and the Same?

Now, within the considerations of Western philosophy, there is an assumption behind this question of whether we can develop a natural science of the mind, and we need to tackle that assumption first. The prior question that we need to address has to do with whether mental states or the mind are identical to brain states. Are mental states, in fact, states of the physical brain, or are they something that can exist independently from the brain? When I remember something, is that a state of my brain, organized and processing in a particular way? When I see something, is that a state of the physical brain, or is it a state of something else?

As you know, there is a deep division on this question. Some people will answer yes, and some people will answer no. First, I want to look at the negative answer.

Descartes was a classic dualist. His idea was that mental states and the brain are two very different kinds of thing. The brain, like the body, is a physical thing, which, as he put it, has extension, position, and mass. The mind does not.

DALAI LAMA: In this view, is there an assertion of a self? Is the self equated with the mind that is so distinguished from the brain? Are they one and the same? Do the philosophers make the assumption with this dualistic theory that the mind is a soul after all?

In this instance, the translator's choice of words moves the discussion into a realm that was not intended by the original query, but was fruitful nonetheless. The Tibetan term gang zag, *translated as "soul," actually means "person" or "individual"; this does not carry the heavy religious and metaphysical connotations of the word* soul *as it is used in English.*

PATRICIA CHURCHLAND: Yes, the soul, the spirit, the mind, are rather interchangeable according to the Cartesian concept.

DALAI LAMA: But are they separate? Do they have separate identities? Do you distinguish the self, the "I," from the mind?

PATRICIA CHURCHLAND: I think from Descartes' point of view, one uses "the mind," "the soul," and "the spirit" interchangeably.

DALAI LAMA: Is there a sense that the soul itself simply sees, or that the soul sees via perception or via awareness that is somehow distinct?

PATRICIA CHURCHLAND: I don't think it is supposed that the soul sees through anything. I think it is assumed that the soul sees, the soul is aware, the soul thinks, the soul reasons, by itself, though in response to impact on the body. . . . Some people might think that there are parts to the soul. But basically, if you were a dualist, you thought that the thing that did the thinking and feeling, the thing that was aware, the thing that was conscious, was the mind or soul.

ALLAN HOBSON: I think it is not just history but that most people believe this now.

PATRICIA CHURCHLAND: Eccles is a contemporary neuroscientist, a Nobel laureate, who believes stoutly in this dualistic theory.

DALAI LAMA: Would you say that most people believe this, but that most scientists do not?

PATRICIA CHURCHLAND: Most neuroscientists and scientists generally do not hold this view. But Eccles is not the only exception. I wanted to mention him to recognize that there are some very knowledgeable people who do hold this view. I think a great many nonscientists, perhaps the vast majority of the population, think along dualistic lines.

ALLAN HOBSON: The essential linkage to what I understand Buddhist tenets to be is that consciousness is held by dualists to be able to exist apart from and in some degree independently from the brain.

A Materialist Critique of Dualism

PATRICIA CHURCHLAND: Remember our question: Are mental states really states of the physical brain? Materialists, also called Physicalists—these terms are interchangeable—would say yes. They hold that there isn't any independent stuff, any kind of substance, any independent thing. There is just the brain, which is organized in ways that we don't yet really quite understand, that produces things like consciousness, memory, and so forth.

I want to say a little bit about what would motivate someone to be a materialist. I think it requires motivation in the first place because feelings, such as sadness, for example, really do seem very different from, say, a piece of bone or flesh in your arm, or a piece of the brain in your head. Activity of cells in the brain seems to be very different from having a feeling of sadness or happiness, or having pain, seeing blue, and so on.

Even Descartes had contemporaries who said that the idea that there should be two very different kinds of things (body and mind) is very problematic. They posed the great question as to how there can be two-way interactions between two such disparate and incommensurable things. How is it that a mind or a soul, being made of completely nonphysical stuff, can interact with something physical? How can it cause something to happen in the brain? If it is supposed to cause us to move, for example, what is the nature of such a transaction? That was what Descartes really struggled with. And he said that the transactions take place in the pineal gland, a little gland in the middle of the brain which we now know has nothing to do with those functions.[1] But Descartes liked the idea that the pineal gland was in the center of the head.

But even that didn't help with the question of what such a transaction would be like. He said that a very fine, very subtle substance interacted with the physical substance. His critics, especially the Roman Catholic priest, Father Malebranche, pointed out that that was not a satisfactory answer. To have a physical effect, there should be a measurable physical cause by this immaterial stuff. And they didn't see how that could happen.

The second argument of Descartes' critics was that reality can be very different from appearances. We know that substances can be made of elements that individually look one way, and behave one way, but when combined, look and behave in different ways. We know the earth looks flat but turns out to be round, and so on. Critics argued that even though our experience seems to be very different from the behavior of brain cells, that doesn't mean they *are* different. Seeming to be different is not in fact evidence for things actually being different. That was a criticism made in the seventeenth century that still persists.

Amongst contemporary scientists, there is more than just negative criticism of dualism. There are positive reasons for believing that dualism is probably false. One of the most important reasons has to do with the dependencies between psychological states and changes in brain states that we can observe and quantify. For example, by administering specific drugs, we can change a person's perceptions, or change their capacity to remember things. These observations suggest that there is a very close relationship between particular chemical substances, particular psychological states, and particular brain states. Indeed, the relation between psychological states and brain states is probably one of identity.

Antonio Damasio and Larry Squire will both talk about brain lesions, where damage to certain parts of the brain interrupts or changes psychological functions, so that people are altered in what they can see or remember, what they experience and how they think. A lesion may change whether they can see in depth, whether they can see color, and so on. These changes seem to be very specific and to be related to specific, localized brain structures. Once more, there is a structural/functional dependency that is very striking—so much so, that it does not appear necessary to postulate some other agency such as a nonphysical soul or mind. It seems as though the only thing necessary is this wonderfully complicated, dazzlingly organized brain.

For another example concerning mind/brain dependencies, consider electrical stimulation. With the brain exposed in a patient undergoing brain surgery, the surgeon may need to test and identify certain brain structures functionally, to map specific sensory or motor areas by eliciting corresponding responses to electrical stimulation. Using fine electrodes and weak electrical current, localized parts of the brain have been explored in this way in many patients. This provides a very striking example of dependencies which have been more exhaustively explored in animals, particularly in primate studies. When a particular part of the brain is stimulated, the patient may find that they can't express certain words that they would like to utter. Or they may unexpectedly experience very specific memories that return to them from the remote past, or they may hear old popular songs.

If there were a soul in there, you might wonder how the electrical current has these effects. Does the soul somehow intervene at the points of stimulation? It doesn't appear likely.

We also know about external and internal accidents affecting brain tissue, with localized or diffuse disorganization and even generalized degeneration of brain tissue, which are associated with a corresponding variety of specific psychological effects. Once more, there is a remarkable correspondence between site and distribution of brain damage and the kind of disruption or degeneration of psychological functions, with loss of perceptual, emotional, judgmental, and behavioral abilities. Loss of visual, language, or memory abilities are among the most instructive.

If it were the case that there is a consciousness independent of the brain, and which can depart from the brain at the time of death, then that consciousness is supposed to carry along with it memories that that person possessed. But when the brain deteriorates and memory correspondingly deteriorates at death, or if the brain degenerates long before death and the memory similarly declines beforehand, how does the soul nevertheless retain memories intact? What is your explanation?[2]

Or suppose that a given brain didn't degenerate prior to death. If memories are encoded in the brain as a result of the way that selected neurons interact and change their shape and establish unique circuits, how can that be recorded, or conveyed to, or carried by the nonphysical soul or mind? How can elaborate ultrastructural physical changes upon which the memory depends, and precise kinds of structural dynamics that the brain undergoes whenever we remember anything, relate to the nonphysical soul? How would such changes be transferred into the soul so that, after death, the soul can retain those memories?

How might it work, say, in patients with brain degeneration, who can't remember where they were born, or what they did yesterday, or who their children are? Would the memories of things that they had experienced ten years ago be preserved in the soul, but be unavailable now? That is implausible.

Each of these distinctive sorts of dependencies seems persuasive on its own, and they appear collectively compelling. Thus, the hypothesis that there is a soul—the dualist hypothesis—is not probable. Actually, it is highly improbable.

None of these lines of evidence shows absolutely that dualism is false. I don't think that we can ever show that for any sort of general supposition in science, or anywhere else, for that matter. But it does, I think, make dualism virtually impossible.

The database for all the several kinds of dependencies is very long and I have provided only a few examples. I think another good example has to do with the structural defects and functional deficits that we observe in brains as a result of genetic errors, or as the result of interference with development. Children who are born after a difficult delivery where the oxygen supply to the brain was cut off may have brains that are very defective. You wouldn't expect that reduction in oxygen supply would bother a soul. If it did, it would also do so in the course of ordinary dying.

Finally, there is the question as to how the idea of the mind or soul fits in with the rest of science. There again, I think, the dualist hypothesis is not very compatible with the rest of established science. Admittedly, no one can be quite sure that established science is true, but it looks like the best thing we've got so far, and like Buddhism, it is subject to correction in the light of evidence.

The dualist hypothesis does not fit very well with evolutionary biology. You have a sequence of increasingly elaborate animal species, and then abruptly, lo and behold, humans appear. Unlike anything previously, they possess a soul. This abrupt endowment is especially unlikely because there are such close similarities in genetic material, brains, and behavior between nonhuman primates and humans, and correspondingly no abrupt discontinuities down the evolutionary ladder.

Finally, I would like to pick up on some of the criticisms that were leveled at Descartes. First, it is hard to see what kind of thing a soul could be. If it does leave the body and brain to become separate and independent, how does it maintain its integrity? How can it be one thing, presumably unpartitioned? Is it not like a distributed fog that merges into other things? And if it has parts, such as a conscious part and a nonconscious part, how do the parts interact? What are the nature of any divisions within the mind or soul?

If a mind is a state of the brain, then we might be able to say quite a lot about the differences between what it is when you are conscious and what it is when you are not conscious. But I don't think that we can see what those state change differences would be in the case of a mind or soul.

And similarly with respect to perception: Quite a lot of progress has been made by neuroscientists and psychologists working together to understand perception in terms of the way the cells in the brain respond and interact, and so forth. There isn't any account of what the mechanisms or process would be like, or how the flow of consciousness or thoughts would proceed. If you do postulate the existence of a soul, there are a great many explanatory difficulties. Moreover, there is no need to postulate a nonphysical mind or soul apart from the brain, because we can account pretty well already for these phenomena in terms of brain properties, dynamic circuitry, electrophysiological properties, etcetera.

I will say one last thing, picking up on Bob Livingston's introduction. Within the last three decades there has been spectacular, unanticipated, almost incredible progress made in the neurosciences. I think that we are living at a very special time when psychological properties can find solid explanations in terms of neurobiological properties.

It is not surprising that understanding about the mind/brain comes so recently, because investigating the brain is a highly technical affair that requires a great deal of specialized technology. When people tried to understand brains or minds before physics and chemistry, before evolutionary biology and molecular biology, before microscopes—including electron microscopes—before electronic computers, and magnetic resonance imaging, they could not get far. Until very recently we did not have either the theoretical foundations or the technical finesse required to investigate these extremely delicate yet comprehensively integrated processes. Now things are really coming together, and although it is only a beginning, it is a very promising beginning.

The Technological Bias of Mind/Brain Metaphors

Robert Livingston adds a coda to Patricia Churchland's presentation, pointing out how rooted in its own fascination with technology, and thus culturally bound, are the various metaphors that science has used to envision the relationship between mind and brain.

ROBERT LIVINGSTON: In a way, Your Holiness, models proposed to account for the correspondences between mind and consciousness on the one hand, and brain mechanisms on the other, historically have consistently followed the contemporary technology. For example, Descartes accounted for this relationship by analogy with hydraulic systems popular then for making elaborate fountain displays, operating clockworks, and animating puppets. Such hydraulic systems were invoked by him to explain nervous system functions in relation to the soul. For him the nerves were hollow tubes containing fluid. He knew that they are arranged in an orderly fashion peripherally and centrally. They responded to stimulation by imparting motion to the fluid. Descartes thus provided the first comprehensive physiological explanation of the nervous system as an automaton. He postulated that the nerves all terminated by opening into the brain ventricles. Their activation would project extremely fine effects, by means of fluid jets directed against the pineal gland. Thus, when a particular part of the body is affected, such jets would ballotte the pineal delicately from that particular direction.

Descartes illustrated this with the picture of a young boy near a flame. The boy is looking at the flame which is quite close to his foot. Descartes writes that the quickness of the fire transmits itself to the skin and the nerves in the skin. And these pass by through the leg and spinal marrow, to the inmost part of the brain, where they debouch into the cerebral ventricule. When nerves in the foot are activated, they direct a fine jet of fluid which strikes the pineal gland from a specific angle. By virtue of the orderly disposition of nerves in the body and their debouchment into the ventricular pool, the fluid jet informs the rational soul in the pineal gland that there is fire in the vicinity of the foot. This is accompanied by conscious perception of heat, reflex withdrawal of the foot, and willed retreat. The pineal commanded rational behavioral responses by perturbing the ventricular chambers and hence the corresponding motor nerves. That is how it also controlled consciousness, sleep, and dreaming, by inflating and conflating the ventricular system. Thus, Descartes understood reflexes and gave them that name by analogy with the geometrically orderly reflection of light from a surface.

In more modern times, it was traditional to speak of brain mechanisms as if they were similar to the telegraph, then later as telephone systems with switchboards, and now the computer. As you heard from Newcomb Greenleaf in the first Mind and Life Conference, computer scientists and cognitive scientists are increasingly using various configurations of solid state physical systems (euphemistically called "neurons") to create decision-making machines, language interpreting machines, and so on. This gives people an increasingly buoyant feeling that they are making progress in understanding the brain.

But I think Western neuroscientists are inclined to believe that there is no model that is entirely appropriate, as yet, for the brain. Living cells possess awe-inspiring capabilities: extremely fine structures that are dynamic and responsive to their near and remote environments with very complicated and intimate structural relations that thrive and are selected as circuits according to nerve impulse traffic. Most importantly, they are organized in ways that provide for local decision making, even within the ultrastructural architecture of a small number of cells. Does that sound reasonable to you?

PATRICIA CHURCHLAND: I don't know whether that is widely held, but I would argue that the brain *is* a kind of computer. It clearly isn't a kind of hydraulic system, or telegraph, or telephone switchboard. But I think it really is a kind of computer. What kind of computer it is, we don't know yet, although we have some reasonable ideas. It is clearly not exactly like the kind one has on a desk. It doesn't have that sort of architecture, and it doesn't use those principles. But it must be the case that the brain is doing computations. It has to if we are to be able to see stereoscopically, for instance. It has to take information from the two eyes, compute the disparity, and then determine depth and what is behind what. Doing that through detailed interactions of neurons is best understood as computing. The exciting thing, I think, is that when we find out what kind of computer it is, that will give us radically new ideas for technology.

There already are powerful and capable network models that are yet too simple to be realistic models of the brain, but they have all the right "smell" about them. Currently, simple models of neural networks can perform interestingly complex computations. They seem to be on the right track as models of brain function. And they promise to become technologically enormously valuable.

3. A Buddhist Response

B. Alan Wallace

By drawing a line between the two camps of dualism and materialism, Dr. Churchland defines the debate in purely Western terms. Buddhism, however, fits comfortably into neither camp. Many of the arguments that have long been leveled at religious thinking by scientists are aimed squarely at the Judeo-Christian tradition, and miss the mark when turned on Buddhism. Some of these issues are addressed in the discussions that evolved over the two days of the conference, but the absence of a shared terminology and the limited time available often left important aspects of Buddhist philosophy and psychology incompletely articulated in the dialogue.

It is not evident, for example, that the Western scientists participating in the conference understood that Tibetan Buddhism explicitly rejects any substantial dualism of mind and matter in which the two are asserted as self-existent things or substances. For example, at one point Patricia Churchland commented, "We don't have the slightest idea of what it would be for a soul to see red, or feel pain, or think interesting thoughts." But here, the "soul" is being treated as a unified "thing"—a concept utterly alien to Buddhism. Buddhist psychology, in contrast, examines individual perceptual and conceptual processes and investigates the causal nexus in which each of them arises.

Serious misconceptions can pass unnoticed when we equate superficially similar constructs that are rooted in fundamentally different systems of philosophy. Tibetan Buddhism has never harbored the notion that our day-to-day mental processes exist independently of the body, as Allan Hobson implies when he formulates the relationship between Buddhism and dualism: "The essential linkage to what I understand Buddhist tenets to be is that consciousness is held by dualists to be able to exist apart from and in some degree independently from the brain." Nor has Buddhism ever drawn anything like the Cartesian distinction between humans and animals. In the Buddhist view, both humans and animals are similarly endowed with consciousness, which functions in dependence upon their physical organisms.

But the lack of congruence between Tibetan Buddhism and dualism is not by any means to say that Buddhism is in agreement with the materialist ideology promoted by Professor Churchland.

A Middle Path between Dualism and Materialism

In order more fully to understand the Tibetan Buddhist perspective concerning the body and mind, let us begin by looking at its view regarding the nature of phenomena in general. The Madhyamaka, or Centrist, view adopted by Tibetan Buddhism at large challenges the assumption that any phenomena that comprise the world of our experience exist as things in themselves. Thus, not only does this view reject the notion that the mind is an inherently existent substance, or thing, but it similarly denies that physical phenomena as we experience them are things in themselves. For this reason, the notion of an absolute, substantial dualism between mind and matter is never entertained.

According to the Madhyamaka view, mental and physical phenomena, as we perceive and conceive them, exist in relation to our perceptions and conceptions. What we perceive is inescapably related to our perceptual modes of observation, and the ways in which we conceive of phenomena are inescapably related to our concepts and languages. This theory is not alien to science. One of the great architects of quantum physics, Werner Heisenberg, declares, for example, "What we observe is not nature in itself but nature exposed to our method of questioning."[1]

In denying the independent self-existence of all the phenomena that make up the world of our experience, the Madhyamaka view departs from both the substantial dualism of Descartes and the substantial monism that seems to be characteristic of modern materialism, or physicalism. The materialism propounded during this conference seems to assert that the real world is composed of physical things-in-themselves, while all mental phenomena are regarded as mere appearances, devoid of any reality. Much is made of this difference between appearances and reality.

The Madhyamaka view also emphasizes the disparity between appearances and reality, but in a radically different way. All the mental and physical phenomena that we experience, it declares, appear as if they existed in and of themselves, utterly independent of our modes or perception and conception. They appear to be things in themselves, but in reality they exist as dependently related events. Their dependence is threefold: (1) phenomena arise in dependence upon preceding causal influences; (2) they exist in dependence upon their own parts and/or attributes; and (3) the phenomena that make up the world of our experience are dependent upon our verbal and conceptual designation of them.

This threefold dependence is not intuitively obvious, for it is concealed by the appearance of phenomena as being self-sufficient and independent of conceptual designation. On the basis of these misleading appearances it is quite natural to think of, or conceptually apprehend, phenomena as self-defining things in themselves. This tendency is known as reification, and according to the Madhyamaka view, this is an inborn delusion that provides the basis for a host of mental afflictions. Reification decontextualizes. It views phenomena without regard to the causal nexus in which they arise, and without regard to the specific means of observation and conceptualization by which they are known. The Madhyamaka, or "Centrist," view is so called because it seeks to avoid the two extremes of reifying phenomena on the one hand, and of denying the existence of phenomena on the other. From this perspective it seems that the above materialist ideology falls to both extremes, by reifying physical phenomena and by denying the existence of mental phenomena.

In the Madhyamaka view, mental events are no more or less real than physical events. In terms of our commonsense experience, differences of kind do exist between physical and mental phenomena. While the former commonly have mass, location, velocity, shape, size, and numerous other physical attributes, these are not generally characteristic of mental phenomena. For example, we do not commonly conceive of the feeling of affection for another person as having mass or location. These physical attributes are no more appropriate to other mental events such as sadness, a recalled image from one's childhood, the visual perception of a rose, or consciousness of any sort. Mental phenomena are, therefore, not regarded as being physical, for the simple reason that they lack many of the attributes that are uniquely characteristic of physical phenomena. Thus, Buddhism has never adopted the materialist principle that regards only physical things as real.

4. The Spectrum of Consciousness

From Gross to Subtle

Patricia Churchland's presentation sparked a discussion in which the Dalai Lama distinguishes between different types of awareness. Buddhist psychology identifies a spectrum of consciousness from gross to subtle, with the gross being equivalent to those aspects of consciousness dependent on the brain that are recognized by Western science. At the far end of the range, subtle consciousness is least dependent on the physical brain.

The very idea of subtle consciousness is problematic for the scientists here, who assume all consciousness to be an emergent property of the physical organization of the brain. Aside from the philosophical problems surrounding causal relationships between nonphysical phenomena and the physical world, there are also problems of a different order, as two disciplines try to communicate without a shared terminology. Where Tibetan Buddhism uses the term subtle *with the technical precision appropriate to an elaborate body of theoretical knowledge and a long tradition of empirical testing of that knowledge, its meaning to the scientists is vague and sometimes contradictory. The problem of subtle consciousness returns again in various forms throughout the conference and proves to be the most stubborn of sticking points in the dialogue.*

DALAI LAMA: There are a great many varieties of awareness and degrees and qualities of consciousness. Some, which are of a grosser nature, are entirely dependent on the brain. In respect to them, unless the brain functions, these grosser mental experiences will not occur.

That is interesting, but I should like to know: What happens when you get down to a subtle level of functioning in the brain—a level of functioning that is very minute in scale and degree—do you understand? The grosser levels of mind, or awareness, are heavily dependent upon the physical brain. If we take that principle farther into fine detail and subject it to analysis in its finest, uttermost detail, the question is: Does *that* level of awareness arise in response to stimulation, or can subtle activation of the brain be generated by a subtle change in the mind, or by something else, perhaps extracorporeal?

I think of awareness as being of two types: conceptual awareness and sensory awareness. It is quite clear that sensory awareness is directly dependent upon the physical components and functioning of the body. Now, turning to conceptual processes, when we think about something, isn't it evident that changes in the body can occur as a result of our thinking? When conceptual processes occur, are they produced by neural processes which give rise to the mental processes, which in turn give rise to further ramifications in the body, or is it otherwise?

Sometimes when we recall something, we get some kind of image; then, depending on that image, we engage in various thoughts. There will certainly be some physical basis, some physical stimulation, that gives rise to that original recollected image.

But isn't it true that on some occasions—seemingly out of the blue, as it were—a thought arises, a mental image comes into awareness, which may have a number of important implications, repercussions, and effects? The question then is: What causes those conceptual events which occur without any cause that we can discern subjectively? Are they elicited by something occurring in the brain, or might they have some other source or origin of stimulus?

PATRICIA CHURCHLAND: It must have a cause in the brain, or at least that's our expectation. Of course nobody has proved that that is inevitably the case, because our investigations of the brain haven't progressed that far. But it is hard to see how it could be otherwise. Not being aware of the cause does not entail that there is no cause. It is just hidden from awareness. People were unaware for a long time of the causes of planetary motion, but it definitely has causes.

ANTONIO DAMASIO: There is one example that might be important at this point. When something appears to be coming out of the blue, one possibility is that it may be triggered by something explicit, but that it appears to us to come out of the blue because our conscious attention is focused elsewhere. We normally are able to focus only on a small fraction of brain events at any given time, and most of what is happening in our brains and also around us is in fact not under the range of our attention. This applies not only to what we see and hear outside, but also to what goes on in our internal thought processes.

There is definitely only a very small range and content of information that we can ever attend to. So something that was not attended to, a stream or ongoing chain of thoughts that you might not be aware of, could suddenly pop up. If your attention had not moved in that direction, then that would appear to come all of a sudden, out of the blue. Yet, it was associated with a quite different chain of thinking that you were not then attending to, and which would still be dependent upon regular brain mechanisms.

DALAI LAMA: Must all mental events, even those that seem to come out of the blue, have physical causes? Is your assertion here based upon a great number of observations—that a great number of mental events certainly do arise as a result of cerebral events? Are you making this generalization because it may be uncomfortable to admit that there could be exceptions? Or, have you established with 100 percent certainty that exceptions do not occur and you know explicitly why?

ANTONIO DAMASIO: As Pat said, there is very little that we have established at 100 percent for anything.

DALAI LAMA: Isn't it the case that you have simply not found any mental events independent of physical events, rather than finding that there are no mental events independent of physical events? That is a subtle but important distinction.

ALLAN HOBSON: Science will never be able to prove the latter assertion. Tomorrow we will talk about dreaming. Dreaming depends explicitly on brain states that used to be considered extremely subtle, with no possible physical explanation in brain activity. At present, I think it is very clear that all aspects of spontaneous thoughts that arise in dreams are related to specific activation of the brain and come from no other source.

DALAI LAMA: According to Buddhist theory, there are some things that belong to subtle consciousness, or subtle mind, that are independent from the body, from the brain. There is no assertion in Buddhism that there is a *thing* called a soul or a *thing* called consciousness, some *thing* that exists independently of the brain. There is no such *thing* existing *independently* of the brain or being *dependent* upon the brain. But rather, consciousness is understood as a multifaceted matrix of events. Some of them are utterly dependent on the brain, and, at the other end of the spectrum, some of them are completely independent of the brain. There is no one thing that is the mind or soul.

I am uncertain about Buddhist philosophy or psychology here in terms of the relation between the brain and the body. Although in the traditional Buddhist context there is no specific reference to brain in respect to conceptual thinking, there is reference to the physical activities, faculties, and organs involved in perception. Vision is understood to be a subtle form of matter which is *in* the eye, but I don't know of a specific reference, apart from the eye, for connections back and forth with the brain.

There is a distinction between sensory awareness and mental awareness, and in terms of mental awareness, there is conceptual as well as nonconceptual mental awareness. And certainly it is a fundamental theory of Buddhism that there is disparity between appearances and reality.

So what criticism do you have of the position I have outlined here?

ALLAN HOBSON: I would like to respond directly to the stated theory. I would say the claim that the part of mentation which is independent of the brain is "subtle" is a function of our ignorance of the subtlety of the brain.

DALAI LAMA: When we speak of mental awareness, it does not always refer only to the subtle awareness. From the time of conception to the time of death, the body is obviously functioning in some way, but when the body ceases to function as a body, there is still a very subtle form of consciousness and that is independent of the body. The fact that the body is able to act as a basis for mental events is dependent on the preexistence of a subtle form of consciousness.

What you call consciousness has its basis in a subtle type of awareness. There is a capacity for awareness, a kind of luminosity which is of the nature of awareness itself, which must arise from a preceding moment of awareness. In other words, there is a continuum of awareness that does not itself arise from the brain. This basic capacity exists right from the initial formation of the conceptus, prior to the formation of the brain itself.

ALLAN HOBSON: Western science would obviously not agree with that part of Buddhist theory. We would assume that conscious awareness arises at some stage during brain development, when there are enough neurons with elaborate enough connections to support conscious activity. We would hold that there is no prior consciousness. Consciousness, therefore, is not infinite in our view. It originates in brains, and it is essentially expandable according to the number of brains that have been sufficiently evolved biologically.

When Does Consciousness Begin?

The discussion then turned to an issue that has received much attention in the West. A determination of when consciousness begins is pivotal in the right-to-life debate that has polarized American society, and there is an active committee of the American Academy of Arts and Sciences that is pondering this problem at present. We know that the basic architecture of the brain is formed during embryogenesis, and that the brain is already remarkably well organized in its basic plan by the beginning of fetal life, after only eight weeks of gestation.

But for all the attention the question has received, there is a startling lack of consensus among the participants at this meeting on even the most basic criteria for determining whether an organism is conscious. The confusion is revealing. The scientific exploration of consciousness in the West is so young that we lack even a definition of consciousness that would allow us to recognize it unequivocally.

DALAI LAMA: At what point in the formation of the fetus do you posit consciousness arising for the first time?

ALLAN HOBSON: It is impossible to say at this point.

ROBERT LIVINGSTON: Biologically it originates gradually, so it is hard to say precisely at what point it is sufficient to meet some particular definition of consciousness, to measure its beginning. The beginnings of biological organization, and probably the beginnings of consciousness, arise asymptotically and in accordance with a biological schedule of neuronal and glial cell divisions, migrations, and elaborations into an embryonic brain, with continuing development throughout gestation and postnatally. The brain at birth weighs about 350 grams. It doubles in volume by six months. It doubles again by about the fourth birthday. Thereafter, it increases by only about ten percent, reaching its maximum around, roughly, the twentieth year. Consciousness certainly begins before birth, but how early is by no means established.

PATRICIA CHURCHLAND: How can you be so sure that consciousness begins before birth?

ROBERT LIVINGSTON: For example, you can condition a baby beginning with the early part of the last trimester and can do so more and more subtly throughout that period.

ALLAN HOBSON: Conditionability doesn't imply consciousness.

ROBERT LIVINGSTON: I would have assumed that there was awareness, consciousness, of a kind of enclosure, sometimes red and sometimes completely black, that moves about, and manifests noises like the mother's heart sounds or the sounds of blood coursing through the placenta. In this dynamic enclosure one can move about against both elastic and contractile forces, change posture, relax, suck one's thumb, and even be aware of outside sounds, such as voices and melodies. I would apply the same criteria that I would apply to an animal to decide whether it was conscious or not. A newborn babe looks around, reaches for the source of a novel sound, responds differentially to its mother's voice, and, when hungry, certainly seeks the breast using olfactory, gustatory, and tactile cues, and lets go of the breast when satiated and goes off to sleep. Do you think the child is not conscious until it has linguistic categories in its head?

ANTONIO DAMASIO: It is a tough question. You can have conditioning, but that's it.

DALAI LAMA: I assume that even a fetus well along in its development in the womb must have some sort of tactile sensation. I suppose it is unlikely that it has auditory or visual sensations.

ROBERT LIVINGSTON: It has auditory ability and, specifically, auditory memory. Even a child that is premature by some weeks can recognize its own mother's voice differentially. It exhibits a change in heart rate and level of alertness when its mother's voice is presented by tape recorder. This means it has stored rather elaborate acoustic memories before birth. And, it does not manifest such responses when the voice of another newborn's mother reads the same text on the tape recorder.

PATRICIA CHURCHLAND: Also, a fetus will jump—I mean when you are carrying it, it will jump to a loud noise. I can tell you. But this does not imply anything about awareness.

DALAI LAMA: It must have some experience and feeling.

PATRICIA CHURCHLAND: Not necessarily. It could be just responding reflexively.

ROBERT LIVINGSTON: But recognition of the mother's voice means the child has learned that voice, recognizes that voice, and pays attention to that particular voice.

ANTONIO DAMASIO: But it doesn't mean it is conscious.

ROBERT LIVINGSTON: You're right. But then, that leaves moot many questions about consciousness in humans and animals. We depend medically mostly on behavioral manifestations. Are we not conscious of those events in life for which we may be open-eyed and fully responsive, even behaving admirably, as in habitual behaviors such as driving a car, but about which events we do not thereafter remember?

LARRY SQUIRE: I think that the neuroscientists would say that just as there is tremendous complexity and subtlety to consciousness, so the brain has every bit as much complexity and subtlety and detail. As one looks into the brain in increasing detail, one sees extraordinary specialization, especially in the primates and in humans. For example, the brain has developed from fifty to one hundred different little areas in the occipital and temporal lobes, all specialized for different aspects of vision. Within each of those areas are millions of neurons and connections. Although one cannot prove the hypothesis in any final way, there is optimism that the brain has the capacity, and the complexity, for all of the subtlety that consciousness displays.

I think neuroscientists would be sympathetic to the viewpoint that consciousness itself can have cause. At the level of organization of whole systems, consciousness itself can roll forward, causing the events that come forth in sequence. But the planning and execution of behavior is all attached to the physical substrate which is moving along with it. There could be little causal events at tiny places in our brain, or larger causal events or states of the brain, or constituents of the brain, that are organized essentially at the level of our thinking—events that are attached to the brain and organized at a level that moves forward as cause.

On Specialization and Adaptation

DALAI LAMA: So, for example, in very deep meditation, when the mind is brought to a fine point of concentration, would you assume that this is associated with certain parts of the brain? Some people are able to concentrate very, very powerfully, and others have a limited capacity for that. Do you understand this to be a result of differences or deficiencies in the cells?

PATRICIA CHURCHLAND: Yes, there are differences in some aspects of the brain just as we expect there to be neuronal differences between people who manifest different musical capacities. Some sing very well, others sing off pitch, others have perfect pitch. It is the same with concentration and other mental talents. And evidently these capacities are similarly subject to improvement and discipline with training. It would be wonderful to discover precisely the nature of the biological differences. It would be very interesting to know what goes on in the brain during meditation, for example, and how that compares with playing tennis or listening to music.

ANTONIO DAMASIO: There could actually be two types of differences. One is one's genetic endowment and original brain development reflected in inborn aptitude or potential skill, such as aptitude for acquiring musical talent or the ability to express oneself well in language. Secondly, there may be differences in how one has learned and how one has developed through education.

Differences in ability, such as the ability to meditate and concentrate very deeply, would become exaggerated by practice based on the original differences. But all that would certainly be ascribable to brain events. There are differences that could occur in different brain regions in a concerted fashion, perhaps augmented or entrained by practice, that could relate to those particular abilities to concentrate the mind on a given point.

At this point Robert Livingston provided an illustration of how the mechanisms of perception adapt, even unconsciously, to compensate for an inborn disability. In children with alternating strabismus, he explained, a muscle weakness prevents the eyes from converging correctly.

ROBERT LIVINGSTON: In these cases, the child automatically learns to depend on one eye at a time, using one eye and then the other alternately. Otherwise he would see two images and not be able to tell which one to rely upon. Such a child can do very skillful things like playing Ping-Pong or basketball, using one eye or the other, without being himself aware of which eye he is using. At the same time, conscious perception must be dealing strictly with information from one eye or the other at any given time. This is indispensable for that child's correct application of consciousness to spatial judgments and behavior. His consciousness must be swiftly and quite reliably switching alternately between the two eyes and also switching control of perception in order to guide body musculature, eye movements, etc. The brain is responsible for this abrupt, instantaneous switching between two elaborate neuronal patterns of distribution, serving perceptual awareness but without self-consciousness being exerted on the control of the switching itself. The switching becomes entirely automatic and unconscious in cases of alternating strabismus.

The Continuity of Subtle Consciousness

Allan Hobson now addressed a direct question to the Dalai Lama, the first to be so boldly stated, and a hush of anticipation fell over the room. Years later, Larry Squire reflected on the lasting impression that the Dalai Lama's response made on him and the others present. His Holiness presents the logical reasoning for the Buddhist position, but also acknowledges the anecdotal nature of his evidence, and echoes the caveat earlier framed by Churchland and Damasio, that science can very rarely claim 100 percent certainty for anything. In so doing, he voices a readiness to expose centuries of sacred tradition to the light of unsparingly rigorous enquiry. It was not so much a concession of a point in the debate as a demonstration of willingness to engage with them on their own terms. Would they meet this challenge with equally open minds?

ALLAN HOBSON: What is the evidence, from your perspective, that subtle aspects of consciousness are independent of brain? That's one question. A second question is: Are you really sure about this?

DALAI LAMA: When this world initially formed, there seem to have been two types of events or entities, one sentient, the other insentient. Rocks and plants, for instance, are examples of nonsentient entities. You see, we usually consider them to have no feelings: no pains and no pleasures. The other type, sentient beings, have awareness, consciousness, pains and pleasures.

But there needs to be a cause for that. If you posit there is no cause for consciousness, then this leads to all sorts of inconsistencies and logical problems. So, the cause is posited, established. It is considered certain.

The initial cause must be an independent consciousness. And on that basis is asserted the theory of continuation of life after death. It is during the interval when one's continuum of awareness departs from one's body at death that the subtle mind, the subtle consciousness, becomes manifest. That continuum connects one life with the next.

At this moment, we are using the sense organs at the grosser level; then when we are dreaming, a deeper level of consciousness manifests itself. Then beyond this there is deep sleep without dreaming, with a still deeper level of consciousness.

On some occasions, people faint. Even when your breath temporarily stops, during that moment, there is a reduced level of consciousness. Consciousness is most reduced late in the course of dying. Even after all physical functions cease, we believe that the "I," or "self," still exists. Similarly, just at the beginning of life, there must be a subtle form of consciousness to account for the emergence of consciousness in the individual.

We must explore further the point at which consciousness enters into a physical location. At conception, the moment when and the site where consciousness interacts with the fertilized egg is something to be discovered, although there are some references to this in the texts. The Buddhist scriptures do deal with it, but I am interested to see what science has to say about this. During this period we believe that without the subtle consciousness, there would be a life beginning without consciousness. If that were the case, then no one could ever recollect experiences from their past life. It is also in terms of Buddhist beliefs relating to this topic that Buddhism expounds its theory of cosmology: how the universe began and how it later degenerates.

Based on this metaphysical reasoning and other arguments, and based on the testimony of individuals who are able to recollect their experiences in past lives very vividly, Buddhists make this claim. I am a practitioner, so based on my own limited experiences, and the experiences of my friends, I cannot say with 100 percent certainty that there is a subtle consciousness.

You scientists don't posit consciousness in the same sense that Buddhists do. At the moment of conception, however, there has to be something that prevents the sperm and egg from simply rotting, and causes it to grow into a human body. When does that occur? Why does that occur?

ANTONIO DAMASIO: Biological properties. . . .

PATRICIA CHURCHLAND: Of the cell and DNA. It is an important problem, but it has an explanation that we now understand. It requires no special forces, no supernatural processes, no ghostly interventions.

DALAI LAMA: We would like to discuss that in the next session. I want to know more, you see. Generally I find in the fields of cosmology, nuclear physics, subatomic physics, then, of course, psychology, neurobiology, these fields, there is a cross-connection to Buddhists' explanations. I want to know your viewpoints. Because through you, major scientific viewpoints may suggest some experiments, maybe something very important. At the same time, there are some Buddhist explanations that may provide a different perspective for the West. So this experience is something very useful to me. Very, very useful.

Ever since we met, I have wanted to make clear an idea that is basic to Buddhism, especially of the Mahayana Buddhist approach. Namely, if strong evidence arises indicating that a given thing exists, then it will be accepted. On the contrary, if there is strong evidence that suggests the absence of such a thing—even certain things that are specifically asserted in the Buddhist canon, the original words of Buddha himself—even then, these words are to be interpreted on the basis of valid evidence, and not to be accepted at their face value. In other words, we do not adhere to the literal meaning of Buddha's words when they are refuted by valid evidence.

Because of this basic Buddhist approach, which is very open, I really want to know your viewpoints. So, if you find from your own scientific perspective any arguments against a particular issue asserted in Buddhism, I would like you to be very frank, because I will learn and benefit from that.

Cosmology and the Origins of Consciousness

The discussion here returns to the origins of consciousness. His Holiness explains the causal logic behind the Buddhist understanding of the origins of consciousness and the role of karma in the formation of the universe. In relation to the Buddhist distinctions between sentient and nonsentient, material and nonmaterial phenomena, Robert Livingston presents a scientific explanation of the biochemical distinctions between organic life and inorganic matter.

PATRICIA CHURCHLAND: One part of the picture that I didn't quite understand, or I guess that I disagree with, is the idea that there were originally two very different things that were created. There were material things and there were nonmaterial things.

DALAI LAMA: My understanding is that by and large Western cosmologists still adhere to some form of the Big Bang theory. The question from the Buddhist view is: What preceded the Big Bang?

ROBERT LIVINGSTON: There are a lot of scientists who think that the time has passed for support of the theory of the Big Bang and that there was not necessarily a Big Bang.

PATRICIA CHURCHLAND: Even if that's true, then all we can say is that we don't know what came before the Big Bang, and it could have been a yet bigger Bang. But I think Western cosmologists would say that we don't have any evidence whatever that there was any nonmaterial stuff. We can see the development of life on our planet starting with amino acids, RNA, and very simple single-celled organisms that didn't have anything like awareness, and the development of multicelled organisms, and finally organisms with nervous systems. By then you find organisms that can see and move and interact. So the conclusion seems to be that the ability to perceive and have awareness and to think, arises out of nervous systems rather than out of some force that preceded the development of nervous systems.

DALAI LAMA: The Buddhist view is that in the external world there are some elements that are material, and some that are nonmaterial. And the fundamental substance, the stuff from which the material universe arises is known as space particles. A portion of space is quantized, to use a modern term; it is particulate, not continuous. Before the formation of the physical universe as we know it, there was only space, but it was quantized. And it was from the quanta, or particles, in space that the other elements arose. This accounts for the physical universe.

But what brought about that process? How did it happen? It is believed that there existed other conditions, or other influences, which were nonmaterial, and these were of the nature of awareness. The actions of sentient beings in the preceding universe somehow modify, or influence, the formation of the natural universe.

PATRICIA CHURCHLAND: But then I want to know *why* you think that. What is the evidence for that?

DALAI LAMA: There are some similarities between Western science and Buddhist philosophy in that neither is dealing with absolutes or 100 percent conviction. In this way we are both faced with options, out on a philosophical limb.

The tradition that evolved in India dealt with many fundamental philosophical issues. We have to account for the existence of matter in the universe. Do we want to say it arises from a cause or no cause?

The first fundamental philosophical question is: How do we determine whether something exists or not? That is the initial question. The factor that determines the existence or nonexistence of something is verifying cognition, or awareness: the awareness that verifies. You have some experience; you saw something, so it exists. That's the final criteria.

Within the range of phenomena that fulfill the criteria of existence, there are two categories: things that undergo dynamic changes and things that are permanent or unchanging. The latter are not necessarily permanent in terms of being eternal, but permanent in terms of not changing. (In Buddhism, not everything that changes is physical.) For the phenomena that undergo change, there should be a reason or cause which makes the change possible. We can see that both the universe and human beings have this nature of changing. Therefore, they depend upon causes and conditions.

When we search for the causes, there are two types: substantial causes and cooperative causes. When you speak of one thing being the substantial cause of another, this means it actually transforms into that entity. For example, what exists inside a seed actually transforms into the sprout that arises from it. The seed would be the substantial cause of the sprout, whereas the fertilizer, moisture, and everything else would be cooperative causes. A farmer, for example, would be a cooperative cause for the arising of the wheat crop, but he didn't enter into the wheat crop as did the seed.

PATRICIA CHURCHLAND: This is a little like Aristotle, who spoke of proximal cause and efficient cause.

DALAI LAMA: So we can look at these phenomena that are subject to change, and we can go back to their beginning and ask: Did this arise in dependence on a cause or in dependence on no cause? If we accept phenomena which demonstrate the nature of arising from cause, and then posit an initial stage where there is no cause, that would be inconsistent and very difficult to accept. How can you say, suddenly, that everything happened without previous cause? There's a logical inconsistency in maintaining that something now shows the nature of being dependent upon cause, while at the same time claiming that initially it had no cause.

In the ancient philosophical treatises in India, there emerged two different philosophical systems, or schools of thought, on this question. One accepted that the original cause had to be something external, such as a God. From the Buddhist perspective, it is logically very uncomfortable to posit God as being the one cause of everything. The problem, then, becomes: What created God? It is the same question.

PATRICIA CHURCHLAND: Good. That was the question I was going to ask you concerning the first awareness.

DALAI LAMA: So when we ask, what is the substantial cause of the material universe way back in the early history of the universe, we trace it back to the space particles which transform into the elements of this manifest universe. And then we can ask whether those space particles have an ultimate beginning. The answer is no. They are beginningless. Where other philosophical systems maintain that the original cause was God, Buddha suggested the alternative that there aren't any ultimate causes. The world is beginningless. Then the question would be: Why is it beginningless? And the answer is, it is just nature. There is no reason. Matter is just matter.

Now we have a problem: What accounts for the evolution of the universe as we know it? What accounts for the loose particles in space forming into the universe that is apparent to us? Why did it go through orderly processes of change? Buddhists would say there is a condition which makes it possible, and we speak of that condition as the awareness of sentient beings.

For example, within the last five billion years, the age of our planet, microorganisms have come into existence roughly two billion years ago, and sentient beings, perhaps during the last billion years. (We call "sentient" all beings that experience the feelings of pain and pleasure.) Especially during the last one billion years then, we see an evolution into more complex organisms. Now we humans are experiencing this world. And there is a relationship between our environment and ourselves, in the sense that we experience pleasure and pain in relation to this environment.

From a Buddhist point of view, we ask: Why do we experience this universe in this relational way? The cause of our experiencing pain and pleasure in this present moment in this particular universe means that we must have contributed something, somewhere, sometime in the past to the evolution of this present situation. It is in this respect that the question of karma enters. In Buddhism, it is held that there were sentient beings in a previous universe who shared continua of consciousness with us in this universe and thereby provided a conscious connection from the previous universe to our own.

ALLAN HOBSON: What is the evidence?

DALAI LAMA: Yes, very difficult, very difficult.

THUPTEN JINPA: I would give an answer to that. When someone builds a house to live in, you have to have a plan. And on the basis of that plan you build the house, and then later you can live in it. In the same manner, sentient beings who produced the evolution of this present universe made this plan, but not necessarily consciously, by leaving karmic imprints within their subtle mental consciousness, which later, when activated, made it possible to influence the formation of this universe.

ALAN WALLACE: We do not need to prove everything. We don't need to argue as to whether this physical universe can be experienced. This is obvious.

DALAI LAMA: That's right. Once again, the criterion for determining whether something is existent is whether it is ascertained by a verifying cognition, which means a cognition that is not mistaken. For example, a flower cannot prove the existence of a tree. Because the flower does not have verifying cognition, it cannot demonstrate the existence of anything else. Experiential awareness is an instrument through which we judge whether something is existent or not. So then the question would be: If we really possess awareness, or cognition, then that must also have causes. Specifically, the awareness must have a substantial cause. Now if the substantial cause of awareness is matter, then why is it that some things have no consciousness, and other things, such as animals, have consciousness?

ROBERT LIVINGSTON: I have a suggestion here, Your Holiness. The thing that is particular about organisms is that they are composed of organic carbon compounds. Carbon has four symmetric valences that facilitate the assembly of long-chain compounds. Simple carbon chain compounds can build spontaneously into enzymes and replicating devices such as ribonucleic acid (RNA). And simple proteins are formed from primordial amino acids. These contribute to the production of more carbon compounds and replicating machinery. Thus deoxyribonucleic acid (DNA) evolves.

Out of a variety of such long-chain compounds come things like membranes. Sheets of conjoined, adherent molecules form membranes, asymmetric on the two sides. By surface tension, these form readily into spherules, presenting barriers which favor differential transport of substances into and out of the spherules. The replicating machinery of DNA, RNA, and enzymes is relatively protected within the spherules. This suggests the beginnings of cells. The machinery that can replicate the whole aggregate, by replicating the machinery itself and the spherule, is favored over less competent systems for survival and continuing replication. The operation of this selection process means that evolution is off and running!

Mechanisms for the replication of molecules increased opportunities for the appearance and reappearance—and persistence, by iterative replication—of specially efficient or adaptive, or otherwise advantageous, compounds, cells, and organisms.

The machinery of cells, which is assembled in a variety of short and long-chain carbon compounds, composed of elements compounded from inorganic matter, is centrally important here as it consists of self-organizing, emergent phenomena. By this time at least, the controlled cascade of electron energy, the earliest beginnings of metabolic machinery, has emerged. Photosynthesis, the process of capturing high-energy electrons to make sugars and amino acids out of carbon dioxide and water and nitrogen, is a self-renewing source of high-to-low energy electrons. Self-replicating systems, organized functionally by trial and error developed subtle metabolic pathways. Ways were "found," or selected, to take electron energies—in ordered sequences and branching pathways—from high to low energy levels in small steps. The essential process was the controlled reduction of electron energy in a staircase of small steps, as in a flowing cascade of water, rather than in an abrupt fall as in a waterfall. The controlled energy decrements are utilized to assemble molecules, to transport substances within cells, to transport products across cell membranes, as in secretion and neurotransmission, to provide motile power to cells, to contract muscles, to control the dance of chromosomes and other aspects of cell division, to control the fusion of cells, as in fertilization, and to operate all the innumerable physiological systems in multicellular organisms.

This controlled electron energy decrement is something not seen in the inorganic world. Something radically new has been added to the universe in this process. In the inorganic world, shifts in electron energy level are brusque, abrupt, jolting, with large, precipitate falls rather than slight decrements in electron energy levels. In living systems, delicate metabolic machinery is disrupted by such precipitate falls in electron energy levels. The kinds of abrupt shift in electron energy that occur in the inorganic world are actually destructive of life. The effects of ionizing radiation, as in radioactive damage to living systems, are of this abrupt kind, characteristic of inorganic electron dynamics. This is precisely why ionizing radiation is so disruptive and damaging to living systems.

The metabolic control of electron energy transfer permits extremely subtle metabolic, anabolic, and catabolic processes such as are involved in the construction of the great heterogeneity of carbon compounds that are encountered in living systems. These processes are manifested in all sentient beings, and, ultimately, in the development and evolution of nervous systems.

ANTONIO DAMASIO: It is possible in fact that certain things came about by processes of self-organization, as Bob suggests. Out of a prior low level of order, certain chance events contribute to an increase in order, which, by selection, is preserved and elaborated, providing new opportunities for the addition of increasing complexity and higher-level order. In this way, by Darwinian selection, higher forms of life emerge. All forms of life are self-organizing, including ourselves.

Presently, we have examples in mathematics and in certain systems that have been simulated in computers that indicate convincingly that there is absolutely no guide, no supervisor, no controller necessary for certain things to happen in an orderly and reasonably predictable way. As Ilya Prigogine has shown, information can be gained without violating the second law of thermodynamics.

And the same thing applies to the way embryogenesis organizes complicated physical structures such as our brain. So, the thing we are questioning is whether you need to go back always to some originating planner or supervisor.

In fact, when Thupten Jinpa had introduced the metaphor of the universe built like a house according to a plan, under the influence of the actions of sentient beings, he emphasized that this was "not necessarily consciously" done. Dr. Damasio's attribution of this influence to "some originating planner or supervisor" tends toward a

Judeo-Christian conception of a creator God and misrepresents what Jinpa had been trying to articulate. To address this, the Dalai Lama adds the following clarification.

DALAI LAMA: Perhaps an understanding of what Buddhist philosophers call the four principles of reason might be helpful in understanding how karma conditions the processes of evolution in the universe, and the role of sentient beings who inhabit the universe. Those four principles are: (1) the principle of dependence, (2) the principle of efficacy, (3) the principle of valid proof, and (4) the principle of reality.[1]

Generally speaking, that philosophy states that the universe arises and evolves in dependence upon karma. But how far, and how deep, does that dependence go? To what extent does the karmic influence act on the origin and evolution of the universe and on the way the universe functions? What are the limits? What are the levels of effect of karma?

And also, in what categories of the four laws would the influence of karma come into effect? To what extent is it through the law of nature, the law of dependence, the law of function, or the law of evidence? I myself am eager to investigate the limits of karma. I doubt that all natural phenomena, such as the orbits of the planets, and how a particular tree may grow, depend upon karma, or are influenced or determined by karma. I suspect that these may be determined or brought about by influences apart from karma.

5. Mapping Brain Functions

The Evidence of Damage to Specific Brain Regions

Antonio R. Damasio, MD

The second presentation of the conference, from Antonio Damasio, provides concrete examples of correspondence between the brain anatomy and functionality. Damage to specific regions of the cerebral cortex proves to have precise and predictable effects on human perception, recognition, and language. Such correspondences are fundamental to the Western scientific understanding that consciousness exists solely as a property of the complex architecture of the connections between neurons.

ANTONIO DAMASIO: I have some concrete evidence for Western science's view on the brain. I have selected a few examples of changes in mind processes in individuals who suffer from localized brain damage. If one of us suffered a stroke, and had thereby lost a particular area of the brain, that would provide an opportunity to study the results of the loss of that part. We would learn what the brain can do in the absence of that particular part. I shall present examples of what the consequences are for persons who lose certain brain areas as opposed to those who lose other areas.

A very important fact is that the surface of the brain, the cerebral cortex, is made up of distinctive areas, each of which has characteristic functions. Cortical areas appear similar on the outside, but when we study them under the microscope, they reveal specialized organization and distinctive neuronal circuits that cooperate with each other. Thus, a million cells here and another million cells there can communicate with each other by means of very specific connections. There is no possibility for any one group to talk to all other cells, but rather specific groups talk to one another and receive responses from other particular groups of cells. The arrangement is exceedingly orderly and systematic: certain strict principles of organization obtain. There are distinctively different patterns of intrinsic organization and external connections for each particular cortical area. There is a distinctive architecture, area by area.

At this point, Dr. Damasio projected a photograph of a building. The left half of the image was essentially a black and white photograph, while the right half appeared in full color.

ANTONIO DAMASIO: What this picture depicts corresponds to the subjective experience that one particular patient described to us, namely, the loss of color vision on the left visual field, with color vision preserved on the right. This is an experience that many patients have. They report loss of the ability to see color in one half of their visual field, left or right, rather in the same way that you can lose color on a television screen if you turn the color knob. You can still see different shapes, and perceive depth, and know whether an object moves. But you cannot see color in half the visual field.

We have studied the brains of many such individuals over several years and found that there is one critical area on the inner side of each hemisphere in the brain, damage to which leads to this kind of half-field color loss. On each side of the brain, this area is responsible for providing colored visual experiences to the visual field on the opposite side. It is important to observe that when you encounter similar damage to cortex elsewhere in the brain, no color loss occurs. We can confirm this localization in living patients using magnetic resonance images (MRI).

Dr. Damasio showed the group a series of MRI brain scans with the fusiform gyrus marked in red, noting that such images are obtained without doing any harm to the patient.

ANTONIO DAMASIO: Once you lose this particular area, and this alone, your visual experience will be normal, except for the loss of color in one half-field. There is no possibility of learning or otherwise restoring the experience of color vision in that affected field. If you produce equivalent cortical damage elsewhere, no such loss will occur. In other words, there is a one-to-one relationship between color loss in the corresponding half-field of vision and this particular area of cortex. This area controls a color-dedicated neuronal system.

We find a similar color vision deficit from removal of the equivalent cortical area in the brain of the monkey. Further, we know that neurons in this region in the monkey are responsive to color presented in the contralateral visual field. In other regions neurons may be responsive to visual shape or motion, but not to color, or only incidently to color, as in passing along color information to this particular area. In this region, therefore, there are neurons which specialize in the color processing necessary for conscious perceptual experience of color.

DALAI LAMA: It is possible to see shape without color, for example, in the case of someone who is color-blind, but is it possible to see color without shape?

ANTONIO DAMASIO: That is an interesting point. If other parts of the visual system were damaged and this particular area were left intact, we might suppose that we could perceive impressions of colors, such as seeing blueness, redness, colors without borders. Yet color perception is characteristically confined to specific shapes: for example, this table, your garment, and so on. When we lose the ability to see shape, we also lose the ability to see light, and we develop a blind field. It is evident that the color-bearing signals traverse most visual pathways but that they contribute to conscious perception of color as an expression of cortical analysis in this location.

Now I want to show you another interesting example of mind changes resulting from a stroke. With the lesion demonstrated here (in the medial occipital temporal region) the patient loses the ability to recognize familiar faces. The patient is unable to recognize the faces of friends, members of one's own family, even one's own reflection in a mirror, nor can the patient recognize the faces of familiar public personae. This deficit in facial recognition is specific to particular cortical areas. It is only when you damage this particular area that this phenomenon occurs; damage elsewhere does not cause this effect.

Something quite interesting happens in patients who lose the ability to recognize familiar faces consciously. They are still capable of exhibiting recognition of which they are entirely unconscious. One way of learning about this is to measure skin conductance in such patients. It is known that normal individuals in the process of recognition generate a dramatic shift in electrical resistance of the skin. If I were to encounter Pat Churchland, my conscious recognition of her would be accompanied by just such a galvanic skin response. If we encounter someone with whom we have had no previous experience, there is no matching recognition and there is no galvanic skin response, or only a small, weak response. A patient who cannot consciously recognize faces still produces a galvanic skin response which occurs only on exposure to familiar faces, signaling an unconscious physiological response that accompanies recognition at some unconscious level. With a specific type of brain lesion, we may be consciously oblivious of recognizing someone while at the same time there will be physiological signs of recognition at a lower level of brain organization, accompanied by a galvanic skin response. This indicates clearly that at some unconscious level I do know that person.

DALAI LAMA: The person in this case fails to recognize a specific face, but does he recognize that it is a human face?

ANTONIO DAMASIO: Absolutely. That is an important point. The person knows not only that it is a human face, but also whether it is a smiling or a sad face, a good-looking or an evil-looking face, old or young, male or female. And yet, the conscious identification of the familiar face is not available. At some deeper level, however, there is a specific, personalized, unconscious discrimination.

Now we are considering levels of consciousness in physiological terms and at the level of neuroanatomy. When a patient has suffered the loss of this particular region of cortex, he cannot muster conscious recognition. The patient is unable to use features that identify that unique individual according to the patient's unique personal experiences—the features that evoke the memoranda that identify that individual. Yet, at a lower level of neuronal processing, the patient generates an electrical skin response that is unequivocal, unmistakable evidence that in fact his brain *knows* that person from prior experience. It is a sure sign of previous encounter and thus a sign of unconscious recognition.

The existence of evidence of this sort strongly reinforces the point that different regions of the brain are making decisions, and are contributing in different ways to the integrative mechanisms that support consciousness.

Categorical Distinctions in Consciousness

Is there or is there not a characteristic difference in brain structures and functions involved when we learn to represent, recognize, and categorize different types of objects?

Let's consider responses to a human face, to an object such as a table, to an animal, or to a flower. We now have very solid evidence to indicate that the brain areas, or systems, specifically concerned with human face processing in conscious experience are entirely distinct in location and response from systems concerned with man-made objects, such as tables or chairs or glasses. And those in turn are distinct, also, from systems concerned with representation, recognition, and categorization of living things such as animals and flowers. Our evidence comes from the study of patients in whom we recognized that damaging one area precludes or eliminates the recognition of animals and flowers and other natural objects, but does allow the patient to respond quite rapidly to a long list of inanimate artifacts; for example, that this is a pointer, that these are glasses, that this is a table, and so on.

We also have patients who lose their ability to recognize, for instance, a manipulable object, which can be held and changed in shape by use of one's hands, and yet they can recognize natural objects. This indicates to us that there are distinctions, to a greater or lesser degree, among different categories of objects that are represented in our brains. The evidence indicates that different categories of stimuli are processed differently in respect to how they are stored in memory and how they are retrieved from memory.

There are, of course, other distinctions. For instance, take our ability to recognize musical sounds and to learn melodies. Musical abilities occupy distinctly different brain regions, and I may say, engage different brain strategies from any other categories of stimuli and brain processing capabilities. You can destroy selectively the ability to recognize melodies and to identify pitch correctly, and yet preserve other categorizing skills.

The ability to learn a motor skill involves entirely separate brain regions from those involved in the ability to recognize faces and is likewise distinctly separate from those involved in learning words. In general, language is quite a separate function, distinguished from the rest of these processes by its special location and functional organization.

Another area in which there is a critical difference has to do with the complexity of an object. For instance, picking up on your Holiness' earlier question, when I look into Larry Squire's face, I instantly know that it is a face, a human face, and masculine. And I could even make a suggestion about what, for instance, he might be like as a person. All of that is far simpler in terms of mental determination than saying that this is the face of a particular individual and still distinct from the further determination that identifies Larry Squire's face as familiar to me. For that particular aspect of brain processing, Larry Squire's unique physiognomy, there are uniquely defining features, united in a particular facial representation and connected with a complicated and unique ensemble of associated experiences.

Evidently, the brain is continually pigeonholing objects into broad but distinctive categories: that a flower is a natural object; that a face is a natural object; that a car is an unnatural, man-made, mechanical object. This prompt broad categorization allows the neuronal representations of these different sorts of objects to be separately distributed in the brain for further processing and further categorization by separate, more specialized, areas of the brain. Following "first echelon" initial categorization and immediate channeling to more specialized brain regions, more refined and specialized brain processes apply additional discriminanda by which the objects are further specified and individuated.

With respect to human faces, there are millions of male faces, millions of male faces with beards, and millions of male faces with beards and glasses. These are obviously easy discriminanda, simple to apply, and they enable initial brain processing to arrive rapidly at useful general categorizations. When it comes to knowing that this face is Larry Squire's, that that object is the Eiffel Tower, that that dog is Lassie, that this flower is a cymbidium from my garden, this requires particularizing strategies in specialized locations that can make use of unique, locally stored information. Eventually, to be recognized as familiar, the individualized objects of whatever category have to be linked to remembrances involving unique personal experiences.

So there are evidently numerous ways by which the brain is able to recognize and distinguish these categories in an almost infinitely complicated universe. Separate and distinct brain systems are engaged in managing successive stages of distinctive kinds of categorization.

My final example shows that brain damage in a different brain system, rather than producing a change in the recognition of faces, or objects, or colors, produces a disturbance in the way the patient attends to different objects and the way he examines the visual field. If we are afflicted by this kind of brain damage, called the Balint syndrome, we see the world as though we were perpetually peering through fog, into a visual field in which parts of objects seem to merge into one another and nothing is clear. In these cases, the patient can never attend to more than one particular part of the visual field at a time. The Balint syndrome is a major perceptual disturbance, but curiously, it permits recognition of faces and colors. The affected individual complains of not seeing things clearly and seeing only bits and pieces at a time. Yet, when a familiar face or image comes into a limited field, they say almost immediately, "Aha! There is so-and-so," and identify a particular person correctly. And when the color red comes into a functioning part of their visual field, they describe seeing the red distinctly.

Again, this points to the tremendous specialization of particular brain structures and operations, some of which process color, others form, others motion, and still others categorize whole entities. These regions put various perceptual and mnemonic components together, presenting consciousness with "whole things" represented as categorized, coherent, integrated recognizable objects.

DALAI LAMA: This deficit involves only damage to the brain, not to the sensory faculties such as the eyeball?

ANTONIO DAMASIO: Exactly. If you lose both eyes, you are obviously blind. But in your inner consciousness, you would not have any disturbance of this sort. If you become blind, you can still think in color. You can still think of people's faces, by type or as particular individuals. And you can still imagine a landscape in a perfectly coherent and dynamic way. If you lose only one eye, you only lose binocularity, but your visual field is nearly as complete as before.

The patients I report as seeing everything in a fog are not suffering sensory disturbances, per se. They are suffering disturbances of higher order integration of perceptual information as processed by different parts of the brain. Although the world is put together for us in consciousness, it is assembled from a distributed, analytical, decision-making apparatus which has a surprisingly marvelous gross, microscopic, and ultrastructural architecture.

Another quick example—I think you will enjoy learning about this example because it is quite stunning. If you lose a particular very small area in the left hemisphere, you find among Westerners who speak Indo-European languages a loss of the ability to read, but not of the ability to write. The patient looks at a text and is able to say, "Well, indeed, there are words here, there are letters, but I don't know what they mean." They just look at the text line by line without being able to read it. But when I ask them to write down a text, which I dictate, they fully comprehend the message and write it out correctly. This is because they are working with a different information system in their brain, one which allows the patient to control the hand and write cogently. If, two minutes later, I ask the patient to read what he has written, he cannot read it. He is absolutely unable to do so.

ROBERT LIVINGSTON: You might be able to remember it, perhaps, but you could not read it.

ANTONIO DAMASIO: Yes. This only happens with damage to the left hemisphere. If you have damage in the same discrete area on the opposite side, on the right side, it will not have this effect.

Now it is extremely interesting that if a person speaks a language that is not phonemically based, with a combination of letters for each sound, but instead a language written with ideograms, where a character corresponds to an object, a word, or an idea, then the patient will not lose his ability to read text after left-sided damage. But the patient will lose the ability to read the ideographic language if he sustains a lesion in the corresponding part of the right hemisphere. Thus, following a lesion in his left hemisphere, a patient would not have a reading problem if he has learned a language in which there are ideograms as opposed to letters representing phonemes. And a person with similar left hemisphere damage is not handicapped from understanding sign language, which is a visual language based on the production and interpretation of signs using hand motions and gestures. Once again, this emphasizes how separate and distinctive are the various streams of brain information processing.

ALLAN HOBSON: How does the area represented in this brain lesion differ from that which is concerned with color perception?

ANTONIO DAMASIO: It is a slightly different lesion in terms of size, and the lesion that causes achromatopsia is localized to the gyrus. It does not involve the depths of the sulcus. However, if the lesion goes into what Dr. Hanna Damasio calls the paraventricular area, then you have alexia, the inability to read. If the lesion is further forward, you get color anomia, in which case colors are perceived but without the patient being able to name them.

The Brain's Representation of Body Awareness

There are patients who have damage to structures in the right hemisphere, as opposed to the left. And when the damage on the right side involves areas in the parietal lobe, and sometimes in the frontal lobe also, something quite peculiar happens in relation to the patient's awareness of his own body. The patient can read, write, talk, and be very articulate, but if there is a pain or paralysis affecting his body; the patient is unable to report, in fact unable to realize, that he has any such a disorder. And, the patient does not suffer the normal anguish and concern that one normally associates with cancer or a paralytic disorder.

I think there is interest on the part of Buddhists to discover that there are areas of the brain that control the processing of body awareness. There are areas in the brain that integrate summary representations of body imagery, in much the same way that there are areas that provide smoothly integrated representations of the visual world. Somehow these parietal and frontal areas are important for us to feel the effects of changes going on in our own bodies.

I would like to suggest that when we distance ourselves from our body in meditation, something happens in our brains that allows this conscious perception of separation to occur. It appears to allow perceptual processing to change connections between areas dedicated to body imagery and the rest of the brain. Something evidently happens that divides consciousness and separates perception of one's own body from perception of other events of which we are conscious. This, of course, would be an entirely normal process, but one that some of us would be unable to achieve without disciplined training. I wish I could do that, to be able to make a switching of connections in a physiologically reversible way, as a Buddhist might in a normally functioning way, as opposed to simply determining neurologically that such a separation can occur as a consequence of brain injury or stroke.

ALAN WALLACE: As interpreter, I am concerned with your terminology. You said when a person with this type of damage to the right hemisphere of the brain becomes subject to cancer or paralysis, the person doesn't suffer. Are you saying that the person does not feel physical pain, or does not feel the mental anguish that normally accompanies that pain or paralysis?

ANTONIO DAMASIO: The person can feel a sensation that he or she actually describes as pain, but without experiencing the usually accompanying suffering.

ALAN WALLACE: You mean mental suffering?

ANTONIO DAMASIO: First of all, the patient looks calm and unconcerned. When asked whether he suffers any anguish, he will ask, "Anguish about what?" You then ask, "Do you feel pain?" In some of them, if there is a pain syndrome, they will say yes.

ROBERT LIVINGSTON: Would you like to say something about personal neglect syndromes?

ANTONIO DAMASIO: There is a condition occurring with particular lesions, especially involving the right hemisphere, although they can also involve the left, in which the patient loses the ability to pay attention to bodily events that are occurring on the other side of the body. So, if I have a brain lesion in my left hemisphere, I may be normally attentive and congruent about what happens on either side of my body and the space around my body. But if the brain lesion is in my right hemisphere, and if, for example, someone comes into my peripheral field of visual from the opposite (left) side, I will not automatically look in that direction and pay attention to that field of view. Likewise, I will be less attentive to sounds, expected or unanticipated, originating on that side of me. This is "neglect." You would call it hemi-neglect, affecting the left side of the patient's body and the left side of the world from the patient's point of view.

Neglect actually affects not only the way we perceive the world but the way we perceive our own minds. This is quite striking. We could do the following experiment with a person who exhibits this sort of neglect: Suppose that such an afflicted person, after sitting here at our table, were asked to close his eyes and recall who is seated around this table. A person with right-sided neglect, for example, would name only the individuals seated on his own right side. When asked, "And who else?" the patient would answer, "There is no one else."

However, if you then ask the patient to imagine sitting at the opposite side of the table and to describe, with his eyes still closed, who is seated at the table, the patient would name only those seated on the right side of his imagined position—the same individuals he was previously unable to identify. And again he would be unable to name those on the left side of his imagined position.

So in the patient's mind, the representation or evocation that can be elicited by mental reorientation is also affected. This is a disturbance that affects not only one's perception but the geometry involved in conscious processing of the recall of perceptual experiences.

DALAI LAMA: Are there cases in which brain damage can be cured, or may the brain in some situations heal, or restore, itself? If some portion is damaged and cannot function properly, might other parts of the brain eventually come in to help perform those functions?

ANTONIO DAMASIO: Substitution is a very interesting point. It is true only for certain abilities and for certain faculties.

DALAI LAMA: Is it impossible for substitution to occur in this condition?

ANTONIO DAMASIO: That's true. In most of these cases, they cannot recover normal function. It is true, for instance, for motor function, that if you become paralyzed, generally you will recover some strength. The reason is that the motor system is organized at many different levels, and if you lose motor function organized at one level, other levels may be able to compensate to a considerable degree.

But for certain abilities, like face recognition or language, these are highly sophisticated, highly corticalized abilities. They are organized asymmetrically in the two hemispheres, with little duplication or overlap between them. Then little or no recovery can occur. For instance, if a very skilled musician loses the ability to perceive pitch or to recognize melodies as a result of a stroke, that skill is lost forever.

We have studied musically talented patients. We have followed the case of a professional opera singer who suffered a stroke in his right hemisphere affecting his acoustic processing systems. His ability to recognize arias, even from operas he had sung on stage for years, was entirely lost. His ability to sing on pitch was also lost. But, he can still go to the piano, with a score in front of him, and play perfectly, without error. There is preservation of his ability to read musical notations and to control his piano playing using hands and feet appropriately. He is able to reproduce music correctly on the piano from musical notations; he is able to match movements appropriate for what is indicated on the musical score. But if he tries to accompany himself at the piano and sing, he cannot perform in that way. He cannot control his own musical instrument, his voice, although he can govern his hands musically in playing the piano.

6. Subliminal Awareness and Memories from Previous Lives

Antonio Damasio's description of patients who show a characteristic electrodermal response although they are unable consciously to recognize familiar faces struck a chord with the Dalai Lama. He compares this, and cases of unexplained affinity for strangers, with the experience of memories surfacing from previous lifetimes.

DALAI LAMA: I think many, if not all of us, have experienced occasions when we meet another person for the first time in this lifetime, and we feel some sense of familiarity, or affinity, with that person. This seems to be a commonplace experience. We may meet another person for the first time and there is no such spontaneous, subjective experience of affinity. In the case of spontaneous affinity experienced on a first encounter, will there be comparable changes in galvanic skin response?

ANTONIO DAMASIO: An excellent point, indeed. This is what we call "mistaken identity" in our research protocols. When we make a mistake in recognition, feeling somehow that we know this person, and in fact it turns out that we actually don't, there is indeed a galvanic skin response exactly like that in valid familiar recognition. This accounts for about 10 percent of all experiences we have when encountering unfamiliar faces. In fact, from the point of view of conscious experience, some of those responses coincide with a feeling of familiarity, and the subject often says something such as: "That's very much like so-and-so, but not quite," or, "That reminds me of. . . ." Or they say, "I have the feeling that I know this person, but I believe it is not true." Or, "I really do not know this person although I have a vague sense of familiarity."

DALAI LAMA: Buddhism, of course, asserts the existence of former and later lives. The way this is understood from a Buddhist perspective is that during one's experience in past lives one meets individuals and these meetings place imprints on one's stream of consciousness. The stream of consciousness is then carried over into this lifetime. There is therefore a subliminal affinity.

LARRY SQUIRE: The neuroscience perspective would be that our memories are relatively imperfect, so that when we encounter a person and believe that there may be a sense of familiarity, we are experiencing some correspondence and possibly some confusion with many other faces that we have encountered in this lifetime.

Indeed, there is a phenomenon in aging individuals that we refer to as "generalization of faces," where faces become increasingly familiar. It produces a feeling of familiarity with more faces than when one was younger because we have had so many experiences with faces. Perhaps our mechanisms for facial analysis are not quite as sharp as formerly. This would make us wonder how, if the memory is retained from a previous life, this sense of familiarity is encountered more frequently in later years when you might expect the retained memories to have diminished.

ANTONIO DAMASIO: Our evidence does not speak to the point made by His Holiness. Our evidence speaks simply to the fact that such misidentifications do exist. And we attempt to explain them on the basis of the correspondence that certain facial features do have. For instance, with an image that one responds to with a sense of familiarity but which does not correspond to a truly familiar identity, that face tends to share certain characteristics in terms of physiognomy with other faces, or with a "target" face.

As the discussion continued, Lewis Judd mentioned two other conditions that might have some bearing on the experience of past life memories: the déjà vu phenomenon and a mental disorder called Capgras syndrome, which is characterized by a subjective certainty that a familiar person has been replaced by an identical double. Here, His Holiness comes back to the subject.

DALAI LAMA: You have the case of a person who, because of a stroke in a particular area of the cortex, can't recognize familiar faces. But even without conscious recognition of that person, there is a change in the electrical skin response equivalent to responses in instances of familiar conscious recognition.

I would like to continue along the same line of thought, but not concerning whether you have a recollection of another person, or whether your memory is impaired. Rather, I would like to address the question of your sense of affinity, closeness, or attraction to an object, such as a flower or dog. In some cases we may look at a dog and feel an attraction to it. Here it is not a question of whether you had met the dog before or not, but simply whether you feel an attraction, which in Buddhist understanding is due to familiarity or habituation from the past. Is this subjective sense of affinity accompanied by the electrical skin response that we discussed?

ANTONIO DAMASIO: Yes, it is. It is more difficult to study those skin responses when you do not have unique stimuli, because you have many more confounding factors. But you can do it. Your sense of affinity toward an object, even if it is not an object you have ever encountered before and therefore is a novel object, can still be explained from the Western neuroscientific point of view, as the result of previous memory stores from this lifetime. You will probably experience affinity in respect to certain categories of objects, for instance, certain flowers, certain people, certain landscapes, because these particular objects elicit affective memory responses in your brain and body. That agreeable state can be described in terms of emotional response, in terms of affect. But the affective memories are evoked in your brain because the objects bring to mind (unconsciously perhaps more so than consciously) previous pleasant experiences with *similar* objects or classes of objects. I should point out that emotions and affect are bodily states that result from brain responses that may involve unconscious as well as conscious experiences.

For instance, in our model, a positive or negative emotion can be referenced to a change in the state of a particular region of the body. In other words, we have states of the body that correspond with happiness. This is distinct from states of the body that correspond to sadness, anxiety, anguish, or anger. So emotions in the Western view can be seen as the consciousness induced by specific brain states as accompanied by perceptions of a variety of body responses to these states.

I think of human beings as brains with large bodies on their backs. Each brain carries its body all the time, a bit like a snail carries its shell. We are more or less aware of carrying this body with us all the time. This neuroscience view is different from the commonplace view of consciousness accompanying *pari passu* perceptions of "the world out there." In effect, we tend to think of ourselves as looking out into the world through the telescopic system of an eye, listening through a near-trumpet of a cochlea in the ear, and feeling the palpable world through a glovelike skin, and so forth. The neuroscience view considers that the brain and consciousness are self-organizing, self-activating systems which are central to all experience. The body is a carapace, as it were. The outside world *and* body are represented in brain.

ALLAN HOBSON: I wanted to clarify two points. I think that the Dalai Lama was interested in the face recognition question. I wanted to be sure that we all understood exactly what the implications of that are.

It seems to me that his interest concerns responses to faces that seem to be familiar, but in fact have not been encountered previously in this lifetime. I could phrase his question in this way:

Could this technique be used to help decide whether or not there are representations in our brains of people whom we may have met in previous lives?

I think that's the question His Holiness would like to answer. I'm not sure whether you could do anything with that. But suppose we took pictures of everyone here. Then we could expose them to the 15th Dalai Lama at some future time and see whether he could tell the difference, by conscious recognition and/or by galvanic skin response, between people that the 14th Dalai Lama had seen and people that he had not. There is probably a way to establish whether the number of recognitions is greater than the ten percent that you ascribe to mistaken recognition due to similarities of facial physiognomy.

ANTONIO DAMASIO: I think, first off, that we earnestly hope it will take a long time for such an experiment to be feasible.

ROBERT LIVINGSTON: And therefore, that we should have repeated reinforcement of the Dalai Lama's experience witnessing our faces over those intervening years!

DALAI LAMA: There are instances where small children recollect their past life very vividly.

ALLAN HOBSON: But what is the evidence that they have recollected correctly and accurately? There must be solid, quantitative, documentary evidence, not simply testimony.

DALAI LAMA: There are specific recent instances in which two girls in India recollected the names of people that they had known in previous lives. They knew the geography and the geographical names. They could recollect their home village of their previous life and call it by name. They also recognized textbooks used in their studies in a previous life, but they couldn't even read the texts in their present lives.

PATRICIA CHURCHLAND: How old were they?

DALAI LAMA: Four or five.

PATRICIA CHURCHLAND: They could have talked to practically anyone in the meantime who could have told them stories of the region and the people.

DALAI LAMA: In this case, both girls now have four parents each. Because their memory is so clear, so convincing, the previous two parents now accept each of them as being also their own child. It was a case where the children recollected people and places precisely from their previous lives—places they had never been to, where their parents had never taken them, nor had their parents told them about them. It was totally out of their experience in the five years of their lifetime. They were able to recognize books that they were very fond of in their previous life. Although they couldn't even read, they were specifically attracted to those particular books.

ANTONIO DAMASIO: A positive response.

DALAI LAMA: I remembered this case because one of the two girls had died in an accident with one part of her brain damaged. I myself, as a Buddhist, can't find an explanation for that small girl. You see, she recognized her previous parents very clearly, and also recognized her own previous books. Yet she cannot read. From the Buddhist viewpoint, this is very difficult to explain.

ANTONIO DAMASIO: So she was taken by people to the place where she lived in her previous life?

DALAI LAMA: That's right.

ANTONIO DAMASIO: And in her previous life she had the accident?

DALAI LAMA: Yes. She was only about fifteen or sixteen when she died in the previous life. The present parents, when they first detected these peculiarities, ignored the issue. The girl said this is not my place, my place is somewhere else which has a different name. The parents of this life didn't take it seriously. They thought she was simply fantasizing. But she insisted continuously. One day, her father told her, "Yes! All right, now show me." The small girl took her father very quietly a few miles away, to her previous home.

PATRICIA CHURCHLAND: It does raise the question of why it doesn't happen more often. If all of us are thought to have souls that come from other lives, then the problem is why each of us does not recollect in great detail an earlier life. I quite honestly don't have any beliefs about my earlier life.

ANTONIO DAMASIO: That's exactly what I was thinking. Why doesn't it happen more often? What kind of education was that girl subject to? Was there anything special in that girl's upbringing that would have made that more likely?

DALAI LAMA: In this particular case, I don't know how to explain it. In her previous life, she was an ordinary girl with no special training. Buddhism generally posits many different degrees of vividness of awareness. One important thing to note is that in her previous life, she had a healthy body but she met with a sudden death. So you see, when death occurs suddenly, if one is in perfect health, one's memories still remain very sharp. From the Buddhist point of view, her recollection of her past life might have some relation to meditative experiences that were acquired in previous lifetimes. Ordinarily, it is difficult to remember one's past life. Such recollections seem to be more vivid when the child is very young, such as two or three, and in some cases even younger.

LEWIS JUDD: Does that recollection attenuate as the child gets older and when they grow up? Does it disappear?

DALAI LAMA: Yes. When the present body is fully formed, the ability to recall past life seems to diminish.

LEWIS JUDD: What would the explanation for that be?

DALAI LAMA: Simply the dependence of the mind upon this body. The mental associations with this life become increasingly dominant. There is a close relationship during the first few years of one's life with the continuum of consciousness from the previous life. But as experiences of this life become more developed and elaborate, they dominate.

It is also possible within this lifetime to enhance the power of the mind, enabling one to reaccess memories from previous lives. Such recollection tends to be more accessible during meditative experiences in the dream state. Once one has accessed memories of previous lives in the dream state, one gradually recalls them in the waking state.

7. Steps toward an Anatomy of Memory

Larry R. Squire, PhD

Larry Squire presents an overview of the current scientific understanding of memory and how it relates to the brain's organization. Functions of memory have been found to be distributed throughout the brain, insofar as memories reactivate the original pathways of the experience remembered. But they are also localized in specific brain structures that control memory, independent of other aspects of awareness. Distinctions are drawn between different types of memory. There is short-term and long-term memory; declarative memory in which we consciously hold a representation of some aspect of experience; and nondeclarative, unconscious memory that manifests, for example, as improved performance in a learned skill.

LARRY SQUIRE: Problems of brain function are so diverse and numerous that there are many thousands of investigators in this country studying brain function. There are many groups studying even a problem as specific as memory from many different points of view. This work is panoramic and proceeding at many different levels of analysis: How do single neurons change? What is the anatomy, and the chemistry, of such changes? How do neurons combine themselves into functional networks? How do these networks change? How is memory organized at levels of behavior and cognition? How is memory stored anywhere in the brain? How many different kinds of memory are there?

One might say that in the neurosciences there are two great problems: there is the problem of the initial organization of connections among nerve cells in the brain, and there is the problem of how these original connections can be altered. Initial connections are concerned in part with the things that Dr. Damasio was talking about. We inherit, when we are born, abilities for many different kinds of functions, for example, to perceive visually, auditorily, somesthetically, and by means of olfaction and taste. Infants do well at localizing sounds and responding to melodies. They can distinguish their own mother's voice from others. They are able to perform considerable coordinated locomotion. And each normal human baby inherits the remarkable capacity for developing language.

We inherit those abilities that have been shaped by selection during millions of years of evolution, a kind of cumulative record or memory of abilities that have favored survival and reproduction among our ancestors.

In addition to initial neuronal connections that we inherit, most animals also have the ability to change perceptions and behavior as a result of experiences in their own lifetimes. Experience can modify the nervous system, and these modifications enable us subsequently to behave differently and think differently, as a consequence of individual experiences. We call that the ability to learn and to remember.

Although we do not understand the physical basis of memory in much detail, most neuroscientists believe that memory is recorded in the brain as changes in the strengths of connections among neurons. Thus, there are networks of neurons that operate initially in a certain way. As a result of experience such networks may exhibit changes in the strength of some connections. Some connections may get stronger, others may diminish in strength. As a consequence of such changes, the network then functions differently.

Memory Reenacts Perception

One can usefully think of memory as an extension of perception. Ordinarily, perception can be transformed into memory. We use the word *encoding* to refer to memory deposition, or memory registration, at the time of learning. We know that encoding depends on many different factors such as the amount of attention that is being paid to an event, how important the event is to that individual, and the extent to which that individual can categorize or otherwise organize the event in relation to preexisting knowledge.

The results of the encoding can be held in storage so that later on, even many years or decades later, it is possible to recall that event. In part, we think recall occurs through reactivating some of the same neuronal patterns that were important when the information was being registered in the first place. So one of the fundamentals that we hold about memory is that it is directly linked to perception. Not only do we have in our brains specialized mechanisms for reading, perceiving music, and discriminating speech, but we also have the ability to retain in memory the consequences of each of these different kinds of perceptual neuronal processing. We believe that memory is stored in a spread-out, neuronally distributed way in the brain and that memories are stored physically in the same areas that were initially involved in processing and analyzing the event that is to be remembered.

Isolating Memory: The Evidence of Damage

One of the surprising discoveries of this century is that despite the fact that memory is closely linked to perception and attention and other intellectual faculties, nevertheless, damage to one or two very specific areas of the brain will cause isolated memory problems. In these cases, memory can be impaired without there being any impairment of self-awareness or intellectual functions. A very tiny part of the brain is involved, whereby damage can cause such specific memory problems.

Dr. Squire presented a magnetic resonance imaging (MRI) brain scan showing details of the hippocampus.

Here we have an image of the brain of a sixty-five-year-old gentleman who developed a significant memory impairment. In this area of the hippocampus is a very small abnormality—the only abnormality that we can detect using MRI images.

He has a severe memory impairment and is basically unable to learn from new day-to-day experiences. He essentially forgets the moments of life experience almost as fast as they occur. His memory for something that happened five minutes ago might be about as good as your or my memory for some not very important event that happened a couple of weeks ago. He also has some loss of memories for things that occurred in the past, especially the recent past. His intelligence is normal; he is fully aware and has an intact sense of himself and full insight relating to his impairment. He has an intact memory of his childhood. He has a normal personality and normal skills for getting along in the world.

Until recently, scientists thought about memory as being one thing. We were inclined to presume that this individual had a problem that affected his memory in a global way. Indeed, there is a sense in which his deficit *is* very pervasive because it affects his ability to learn words, to retain conversations, to remember faces, verbal and nonverbal things, spatial layouts, and so forth. His life is terribly disabled, as you can appreciate, because of this memory deficit. Nevertheless, this individual, and other persons like him, have entirely preserved abilities for some other kinds of learning and memory.

DALAI LAMA: Is that person's long-term memory of events before the disability struck impaired?

LARRY SQUIRE: That is also affected, especially for recent years prior to onset of the disability.

DALAI LAMA: Might a person recollect everything clearly up to a stroke or other impairment?

LARRY SQUIRE: That will often happen, but in cases of specific impairment of memory, it is quite common to suffer losses of memory for events that occurred prior to the onset of the disability. We call that retrograde memory loss.

DALAI LAMA: Isn't it possible for some patients to remember things from the distant past without being able to recall recent events?

LARRY SQUIRE: Roughly speaking, that's what we are talking about here. The main deficit is in the ability to store new information. You are particularly able to retrieve older memories of your distant past.

DALAI LAMA: Something that happened five years ago, before the memory deficit began, might be lost, but prior to that time everything would be perfectly recorded and still recallable?

LARRY SQUIRE: That's exactly right. How far back in time the impairment goes and how dense the memory loss is will depend upon the severity of the brain defect.

DALAI LAMA: But does it happen, on occasion, that the memory up to the time of disease onset remains completely unimpaired, leaving memory deficit only for events after the onset of the disease?

LARRY SQUIRE: It would be unusual to have previous memory complete up to the second. People who have auto accidents may have memories that are accurate up to one or two minutes before the accident, and may be very bad afterward. This individual, however, has difficulty remembering events that happened as long ago as twenty-five years prior to the time his brain became affected.

ANTONIO DAMASIO: Severity really reflects having different kinds of anatomical parts damaged in terms of the retrograde amnesia. In virtually all patients, there are some kinds of memory loss for some time before the accident took place. And in some, if there is a certain type of damage, the loss can be tremendous and can affect retrograde memory for decades.

We have one patient whom we have studied for many years, an individual whom Larry Squire has studied also, who has a retrograde memory loss for virtually fifty of the sixty years of his life. There is very, very little memory left at the level of unique events, although he knows something about his life generally.

ROBERT LIVINGSTON: I would like to know, in this case, whether the hippocampus was in your opinion quite intact prior to the stroke. I ask because otherwise a person might have been an alcoholic or have degeneration relating to atherosclerosis and be deteriorating as far as memory is concerned well prior to the stroke, but perhaps family and physicians didn't pay much attention to that until a stroke wiped out new memory storage capacity.

LARRY SQUIRE: We are confining ourselves to cases where impairment came on in a single day. In those patients, if the damage is severe enough, they may lose memory access for perhaps ten or fifteen years prior to the disorder. On the other hand, if the damage is less severe, there may be memory impairment for only a year prior to the disorder. All of these patients that I am talking about here have good, intact memories of their early life, say the first fifteen or twenty years.

DALAI LAMA: Does this show that there are physically different mechanisms for storing and recalling these different periods of memory?

LARRY SQUIRE: Yes. The implication is that the hippocampus is important for storing new memories and for retrieving recent memories. But as memories grow older, something changes in the brain, so that they don't require this structure for retrieval any more.

DALAI LAMA: Is it possible that the storage facilities in the brain for the memories are unimpaired, but that the retrieval system is interfered with? In such a case, might you have the memories cached, so to speak, but not be able to get at them?

LARRY SQUIRE: That is an important question, and often asked. We think that this represents an impairment in storage. This is in part because we have patients who have a memory impairment for a short time, only for one day, for example. And when such patients recover function in these brain sites, they never recover memory for that lost day. It is as though those events were never recorded.

Different Types of Memory

DALAI LAMA: In Buddhist psychology, we speak of memory in terms of retention and storage. Could some cases involve the capacity for retention, but not for the storing of memory?

PATRICIA CHURCHLAND: You don't mean retrieval? Or short-term as compared with long-term memory?

ALAN WALLACE: No. Retention is the initial imprinting process, whereas the subsequent storage is deeper and for a longer term. However, one Tibetan term is used for both processes, imprinting and storage.

LARRY SQUIRE: Western neuroscience makes a distinction between what we call short-term memory and long-term memory. These amnesic patients with memory impairment have intact short-term memory. That is, they can repeat back a short sentence and they can understand the meaning of such a sentence. Thus, they have a capable memory for a short period of time.

PATRICIA CHURCHLAND: How long is short-term memory?

LARRY SQUIRE: It can last from many seconds to a few minutes. If you say, "Four, three, six . . . four, three, six . . . four, three, six . . ." at intervals over a period of fifteen minutes, an amnesic patient can respond by repeating it correctly whenever asked. If you leave and come back in a few minutes to ask what that number was, he will probably be able to respond correctly, "Four, three, six." But then, if you distract him to attend to a different task and ask him again what was that number, he will have no recollection of it.

So the brain makes a distinction between a short-term and long-term memory, and more recently it has become clear that another and perhaps deeper distinction is very important. This is the distinction we would make between declarative, or conscious, memory and nondeclarative, or unconscious, memory.

Amnesic patients have a problem only in acquiring declarative memories, which form the basis later for conscious recollection of past events. Many memories are like that: our memory for conversation, our ability to recall a specific face. But there are other kinds of nondeclarative memories of which we remain generally unconscious.

Dr. Squire demonstrated two tests for the acquisition of nondeclarative memories. The first consisted of a list of uncommon English words printed as though seen in a mirror.

One can learn through practice the skill for reading these words very quickly. Amnesic patients have a completely normal ability to do that. In general, skills are abilities that depend on memories resulting from experience. But skill acquisition does not involve experiences about which we have very much conscious knowledge. We don't have much conscious knowledge about reading, or tennis service, for example. One doesn't know consciously what one does exactly even though one has these skills; one simply does them. And the ability to improve at skills such as reading this kind of text can occur in the absence of the hippocampus, and despite otherwise severe memory impairment.

The second example consisted of a list of words, followed by another list of the same words truncated to the first three letters of each.

Scientists call the second example "priming," as in priming a pump, a useful metaphor for the process. You present a list of words to the subject, and then you display three-letter beginnings of those words. You say to the subject: "This is not a memory test. This is a puzzle. What I would like you to do is to complete each stem or fragment to form the first word that pops into your mind." For normal subjects, there is a very high probability that the words they think of will be the words that they previously saw. And a memory-impaired patient will have the exact same ability and tendency to say those words.

The normal subject will say in discussion afterward, "Oh, that's one of those words you just showed me." He will have a tendency to complete the word and also to remember that it was a word from the list. The memory-impaired patient will produce the correct word, but will not have any conscious recollection that the word had been presented before.

We now think of memory as being composed of many different components. Clearly, we have conscious memory. This kind of memory depends on the integrity of the hippocampus. Conscious memory is concerned with cognition. It is concerned with making representations concerning the external world, about what occurred, and the relationships among events in the world.

Unconscious, nondeclarative memories are not concerned with representations of the external world. They are concerned with behavior, with priming, with improving our ability to correctly identify objects. In these instances, performance improves as a result of experience. But as we have shown, to maintain these kinds of skills, there is no necessity to retain recollections of the past. We simply adapt to the performance requirements for the skill. We simply change our behavior in the course of acquiring skills.

We consider that nondeclarative memories are relatively primitive with respect to evolution. For example, snails and other simple animals have some of these abilities to change behaviorally, to develop skills, as it were. If you touch a turtle on the head many times, it will keep its head inside the shell. The same is true for many snails. The animal simply changes its behavior a result of experience. But we think this is very different from consciously recollecting. There are unconscious abilities to change behavior, such as the abilities to improve the perception of something seen recently or to develop a new movement, which are distinct from the abilities to consciously recollect different occasions of encounters and experiences. And there is one particular brain system, the hippocampus, which is essential for conscious kinds of memory, but not for developing sensory and motor skills.

One of the things that makes this an exciting time to study memory is that we can also study many of these phenomena in animals, including conscious memory in the monkey.

In one such study, the animal sees an object which he must push aside to disclose a raisin reward. Then one minute later, the animal is presented two objects. The monkey will now find a raisin only if he moves aside the new, unfamiliar object. By its performance, the animal informs us that it has recognized the original object as familiar so as to be able to select the other object. Brain lesions in the hippocampus, exactly similar to the patients' lesions, interfere with performance of this task.

One of the important goals of the neurosciences is to understand the significance of brain connections, not only of what is connected to what, but what purposes are served by given connections. Even for something as simple as perceiving a drinking glass on a table, we know now that the brain has two distinctive areas that are each important for that ability. One area, in the inferior aspect of the temporal lobe, is important for identifying what the objects are; the other area, in the parietal lobe, is important for identifying where they are in relation to other objects in space. This dissects perceptual processes in very fundamental ways.

Now, if perception is to be transformed into memory, there need to be connections down to the hippocampus, through stations[1] in the medial aspect of the temporal lobe. All of these must be functional. So, later on, in order to reactivate the representation of the drinking glass on the table, there must have been activity through these circuits at the time of the original perception. These same circuits are thereafter critical for the ability to recollect conscious memories.

8. Brain Control of Sleeping and Dreaming States

J. Allan Hobson, MD

Neuroscience and Tibetan Buddhism have both devoted a great deal of attention to sleep and dream states, and there is a clear overlap of interests in this area. In fact, the topic generated so much interest on both sides that it was to become the focus of the fourth Mind and Life Conference.[1]

Allan Hobson provides a general tour of current scientific knowledge on sleep and dream states, and the implications of these findings for our philosophical understanding of consciousness. Most significantly, consciousness is understood to be a natural condition of the activated brain, and most of the brain activity associated with various states, whether waking, dreaming, or in deep sleep, is generated internally by the brain itself, rather than being driven by sensory input.

Control of the regular cycles of sleeping, dreaming, and waking states is governed by reciprocal systems of neurotransmitters. Because of their role in controlling relaxation, these chemical systems may prove to play a role in meditation also.

Lucid dreaming, in which the subject is conscious of dreaming even as the dream occurs, is another area of particular interest, having only recently been recognized, let alone studied, in Western science. Tibetan Buddhism has a long tradition of dream yoga which includes training in lucid dreaming.

ALLAN HOBSON: I plan to identify some important themes in the science of sleep and dreams, and to orient the discussion toward the clinical and philosophical implications of this work.

Those implications include the following: I think we can now conclude that good sleep and good waking are reciprocally related, as Buddhist philosophy has long held. More particularly, I think we can identify within the brain some of the structural bases of these reciprocal relations. Further, I think we can help understand how some of the practices that the Buddhist tradition has developed actually work in relation to specific brain mechanisms. That is particularly exciting with respect to future experimental prospects.

Obviously, in the West we also believe that sleep is important to health, and to the degree that we can help people learn to sleep without pills, we will be in a much stronger position. One of the great afflictions of the West now is the overuse of sleeping pills. So behavioral techniques, such as those Buddhists have developed, that can be used to help people achieve peace of mind and good sleep are very welcome.

The second point about the clinical implications has to do with dreaming itself. I think we can now objectively identify the brain states associated with dreaming and help people, if they so wish, to have access to their dreams. In other words, we are now in a position more easily to control dreaming, if we want to. I will discuss dream control later.

There are a number of other interesting topics that are of great importance to philosophy and to clinical practice. Philosophically, these issues really have to do with the relationship of brain activity to consciousness. Perhaps the most important general conclusion coming from sleep research is that consciousness in all its varieties is specifically related to particular brain states. Consciousness seems to be a natural condition of the activated brain. Furthermore, most of the activity of consciousness is internally generated. In other words, the brain contains its own mechanisms for creating information, and most of brain activity is concerned with processing its own information, input from the outside world being relatively modest. Previously, in the West, models of brain function were largely input driven. Now, I think we see that the brain has highly organized spontaneous activity, such that the main rudiments of consciousness are inherent, self-organized, and built into the system.

So I hope these three themes—the clinical implications for sleep and its relation to waking, the nature of dreams and our capacity to access them, and the implications for theories of consciousness will be discussed following my presentation.

In such a brief time I can't possibly cover the wealth of material now known about these issues, so I will review five major conclusions.

Measuring Sleep and Dream Cycles

The first is that sleep can be objectively measured. We can, for example, measure activity of the brain, we can measure activity of muscles, we can measure activity of the eyes. When we make these three measurements, we can clearly distinguish three states: waking, sleeping, and sleeping with dreams. Furthermore, we can identify these three states in another way, simply by observing behavior, and using either photography or time-lapse video to keep track of changes in body posture that are associated with the underlying changes in brain state responsible for the shift from sleeping to dreaming. In time-lapse video, a picture is taken at regular intervals, typically every seven and a half minutes, using a video camera instead of a still camera to record a history of postural shifts throughout the night.

We have found that sleepers never make fewer than fourteen major body shifts during the night. They shift from right side to left side to right side to left side during six to eight hours of sleep. And when sleep becomes less sound, they make more movements. With respect to some of your traditions concerning sleeping on the right side,[2] the real issue is, can a trained subject suppress body movement shifting and still have normal sleep? This is a very simple experiment which could be performed with very simple techniques such as time-lapse video, which is easily portable.

Using measurements of the brain, eye, and muscles, we can identify three states: waking, sleeping without dreams, and sleeping with dreams. Nondreaming and dreaming sleep alternate in a regular cycle, which lasts about ninety minutes, with about the first sixty to seventy minutes being nondreaming sleep, the last fifteen or twenty minutes being dreaming. This means that in a night of six to eight hours of sleep, you are going to have four or five dream periods, each lasting fifteen to twenty minutes, or longer. This means that in any given night of sleep, we have as much as two hours of dreaming, distributed at regular intervals throughout the night. There is a lot of time spent dreaming.

DALAI LAMA: Are there any differences due to age?

ALLAN HOBSON: Yes. Sleep generally becomes shorter and shallower with age. And with aging, there is a slight decrease in the amount of time spent in the state of sleep associated with dreaming. The cycle duration remains fixed throughout the life span, once we are adult. Babies have a shorter cycle, and the cycle lengthens as the brain increases in size. The newborn infant has about four times as much dream-state sleeping as the adult does, which is a very interesting point to consider with respect to developmental issues.

DALAI LAMA: Is there any biological evidence to determine when the infant starts dreaming?

ALLAN HOBSON: Biological evidence could never prove when someone starts dreaming because dreaming is a psychological experience, but we know that REM sleep begins in utero. (Here we must be dualists with respect to language. There is an important distinction between language dualism and deep philosophical dualism.)

The three physiological phases that are associated with waking, sleeping without dreams, and dreaming, are characterized by measuring brain wave activity and that of the eyes and muscles. The stage of sleep associated with dreaming is called "rapid eye movement" or REM sleep because the eyes move around very dramatically. At the same time, muscle activity is blocked. Our muscles are paralyzed except for the eye muscles. And the brain is activated. This phase of sleep is clearly identifiable in the human fetus as early as thirty weeks of gestation, and it probably begins at least ten weeks earlier in a more primordial form. So as early as at twenty weeks of gestation, the brain has organized and differentiated sufficiently to generate this kind of alternation between brain states. But the dream state is much more prominent in the fetus even than in the newborn, so that at thirty weeks of gestation, the estimates are that this state occupies about 90 percent of all the time.

Neuronal Controls of Sleeping, Dreaming, and Waking

We know that sleep is organized as an alternating cycle. The next question for the Western scientist is: How is the cycle organized within the brain? We know that it is controlled by brain structures localized in part of the brain stem. The brain stem looks somewhat like the base of a flower. It connects the stem of the flower to the blossom just as in a lotus plant. The bulb of the flower, by this analogy, is the brain stem. This small but very important part of the brain, between the spinal cord and the rest of the forebrain, supports our conscious activities. Neuronal machinery which controls the alternation among sleep stages and wakefulness is located in a small region of the brain stem called the pons or bridge.

This location is obviously strategic. It can control inputs and outputs for the whole body. It can control activity throughout the whole upper brain, the forebrain, which we consider to be the organ of consciousness. Our third point, then, is that this regular alternation in sleep is controlled by the brain stem.

The obvious next question is: How does the brain stem do this? Within this pontine region are two populations of nerve cells that have distinctive chemical signatures. One is the neuronal population that supports the waking state, and we suppose it to be responsible for arousal and even anxiety. Its chemical signaling involves the release of amino acids, hence it is known as an aminergic system. When this system is very active, we are very alert, but we may also become too alert. We may become anxiously alert.

It is the proper regulation of this system that I think constitutes one of the goals of Buddhist meditative practices. And this same system is obviously of great significance also to Western medicine. This is so not only because this population of aminergic neurons controls arousal, wakefulness, alertness, and anxiety, but also because outputs of this system affect vital functions like breathing, blood pressure, and other visceral as well as cerebral contributions to our experiences.

This aminergic system in the pontine part of the brain stem is also involved in energy regulation and energy flow, and probably also with aggressive behavior. I think it is a key to understanding a number of very important aspects of human life. Moreover, it seems to play a significant role in many functions of prime importance in Buddhist thought and training.

DALAI LAMA: Are such emotions as aggression, love, and attachment also associated with that part of the brain?

ALLAN HOBSON: Not specifically, no. But, as part of the general continuum of activation, other forebrain structures will be engaged, which will then, according to external inputs, govern the emotional state of the individual. The emotional system is rather farther forward in the limbic part of the forebrain, including the hippocampus, which was discussed yesterday. This brain stem activation site is not a specific system for the control of emotions. It is a specific system for controlling the level of arousal of the individual as a whole, which thereby affects other systems, including emotional controls.

The aminergic system is one group of neurons in this critical region of the brain stem. The other group of nerve cells in this same region is called a cholinergic system because it's chemical signature—its neurotransmitter—is acetylcholine. We can identify these two neuronal populations in the pontine brain stem, localize their cells precisely, determine their major connections, their chemical neurotransmitters, and record their patterns of electrical activity.

The cholinergic system is apparently held in restraint by the aminergic system. Thus, when the aminergic system is functioning at a high level, the cholinergic system is functioning at a reciprocally relatively low level. That is the situation in the waking state. As we go to sleep, the aminergic system decreases in its activity, and the cholinergic system becomes relatively more active. The cholinergic system becomes progressively more active throughout the period of deep sleep without dreams. Ultimately these two neuron populations become radically differentiated: the adrenergic system shuts off completely, and the cholinergic system reaches its highest level of activity just when you enter the dream state. Activation of the cholinergic cells generates signals that contribute to eye movements, to inhibition of muscle tone, and to activation of the forebrain.

These reciprocal shifts of functional states can probably also be influenced through meditative practice.

DALAI LAMA: Are you indicating that in the dreaming state you are even more relaxed than in the nondreaming state?

ALLAN HOBSON: It is a paradox, because the muscles are completely paralyzed. To speak of relaxation in this case is misleading. The muscles are actively suppressed, or inhibited. But the upper brain, the forebrain, is very active electrically. In contrast to the waking state, this electrically active brain in the dreaming state is chemically distinctly different because of the shift in the neurotransmitter ratios. The dreaming brain is very highly cholinergic, the waking brain is very highly aminergic, while in each of these states the forebrain is highly electrically activated. We believe that this is very important for understanding the differences between the waking state and the dreaming state.

So we now know that sleep is organized into a succession of states. We can identify and measure distinctive sleep states. We know that the brain stem controls the succession of waking/sleeping/dreaming states. And we know that the brain stem controls that succession of states by altering the production of specific neurotransmitters which are represented in two reciprocal systems of neuronal control.

The fifth and final point I want to make about the science of sleep is that we have tested this theory by making microinjections of very small amounts of chemicals into specific, localized regions of the brain stem of experimental animals. By this means, we can control the overall brain states of wakefulness and sleep. In other words, by imitating the activation by acetylcholine in very specific, localized parts of the brain stem, we can convert the whole brain from the waking state to REM sleep almost immediately and keep it there for many hours. If we put the same chemical, acetylcholine, into another part of the brain of experimental animals, we can produce waking. The differentiation of these control systems is specific and precise.

So we have obtained experimental control of the state of sleep in animals. To some extent, similar experiments have been replicated in humans. Obviously, we don't inject chemicals directly into the brain stem in humans. We use human subjects to measure states of sleep and wakefulness and to obtain reports relating to conscious experiences. We use animal studies to investigate what's going on neuronally within the brain stem during different sleep and wakefulness states. All mammals share identical organization of alternating states of wakefulness, deep sleep, and REM sleep (presumably dreaming) behavior. They all have obvious waking states complete with apparent awareness and interactive behavior with the environment. Such waking states alternate with slow-wave sleep, which lacks rapid eye movements and is associated with high-amplitude, low-frequency electrical activity throughout the brain. These slow-wave sleep states cycle regularly into the kind of sleep state associated with globally inhibited body movements except for rapid eye movements, specifically accompanied by low-amplitude, high-frequency electrical activity throughout the brain. In this latter state, all of the objective phenomena are equivalent to the state that in humans is identified by subjective testimony of dreaming.

In humans and other mammals we see a complete suppression of muscle tone during REM sleep, so the motor output is actively inhibited. Otherwise our dream states might be accompanied by our getting up and running around—still asleep—acting out our dreams. Our dreams are typically characterized by the hallucination of movements by ourselves and among other animate and dynamic things. That's because the upper brain, the forebrain, is actually generating elaborate visual and motor patterns which are not allowed to be acted out by our muscles, perforce the general inhibitory control exerted by brain stem mechanisms. Only the eye muscles are permitted to express this internal sensorimotor dreaming state.

Meanwhile, during REM sleep, the brain is electrically activated, even more so than in quiet waking. The brain is intensely internally activated: hence we imagine that the dream arises because the manifestly activated brain is actively processing signals that would ordinarily be associated subjectively with direct, vivid experiences and outgoing behavior. We hallucinate the experiences and the inhibited behavior as if it were not inhibited. And that is our dream!

DALAI LAMA: What accounts for the rapid eye movements? The rest of the muscles of the body are paralyzed in the dreaming state, and yet the muscles associated with eye movements are not. Why is that?

ALLAN HOBSON: The answer is that the eye muscle system is a very different sort of motor system from the skeletal muscle system.

Most of the skeletal muscle system is engaged in maintaining posture against gravity, and that system is obliged to use a lot of tonic inhibition. The eye muscles don't have to do that. The eyes are essentially weightless, their specific gravity is about equivalent to their surroundings in the orbit. The activity of the eyes is to sweep about swiftly and relatively effortlessly. Because they work with straight beams of light, often from very remote objects, they have an enormous leverage and target speed as well as accuracy in relation to the visualized world. The eyes are never completely static and they don't have to work against gravity.

Secondly, the eye motor nuclei which control eye movements lie upstream in the brain stem, forward of aminergic and cholinergic state control systems in the pontine brain stem. All other direct motor systems, for face and lower muscles, lie within the pons or below the pons along the lower brain stem and spinal cord. Hence, the eye motor nuclei are most anterior of all direct muscle control mechanisms. They are relatively so far forward of other direct muscle control systems that they are really a part of the forebrain systems that serve conscious experiences and higher mental life.

A third observation is that moving the eyes does not contribute to skeletal motions and other bodily effects that might have the consequence of jarring us awake.

ANTONIO DAMASIO: Accepting your idea that we have an active suppression of antigravity muscles in order to keep us from moving around, it is clear that the eye muscles would not do that. But do you ever see a lot of movement of the facial muscles, which are controlled only slightly lower, in the lower pontine and bulbar brain stem? They would not have much skeletal motion effects except by way of jaw movements. Moving muscles of the face presumably wouldn't have the consequence of waking you up.

ALLAN HOBSON: You don't see a lot of facial muscular movement, but you do see some. In the human infant, the facial expressions in REM sleep are particularly visible and dramatic, and quite charming. There is automatic smiling that is produced during rapid eye movement sleep. If you watch a mother nursing a baby, you will often see the baby, when he or she becomes satisfied with milk, begin to close the eyes as the sucking movements become very regular and rhythmic. Shortly, the eyes start to move actively, and the baby has gone into rapid eye movement (REM) sleep. Then you see dramatic, spontaneous facial expressions, especially smiling. And the mother believes the baby is sending her a message of contentment and happiness. You can observe this and talk to mothers about this, and they will uniformly interpret the baby's behavior as meaningfully related to their generosity in giving the baby nourishment. That's quite an important ethological concept, this intergenerational signaling system that is so adaptive and useful in contributing reinforcement to the mother for this uniquely mammalian, altruistic feeding behavior.

DALAI LAMA: Dogs make limb movements, and on occasion when a person has a rough dream, like a nightmare, the arms may flail about.

ALLAN HOBSON: Yes, but then it is not a nightmare. There are two kinds of frightening experiences that occur in sleep. Bad dreams, or nightmares, in which you imagine a scenario and frightening things happen to you may occur in REM sleep. They can erupt, which usually results in the subject's spontaneous waking. There is another kind of night terror, which tends to occur in nondreaming sleep but not in REM sleep. This is a purely emotional experience, lacking the associated hallucinatory activity that accompanies dreaming. It is this night terror that may be accompanied by flailing limb movements.

There are some human subjects in whom the brain stem fails to inhibit the skeletal muscles. When they have REM sleep, they act out their dreams. This is a very dangerous brain stem defect.

Now let's look at the way the sleep cycle is organized. First, as the night progresses, come phases of sleep associated with little or no mental activity. Next are periods of sleep associated with dreaming.

The dream periods tend to become longer as the night progresses, lasting thirty, forty, even fifty minutes at a time. So the best time to obtain dream reports is in the early morning when these dreaming periods are quite long.

The cyclic alternation of non-REM and REM sleep is very regular. This suggests that the brain stem neuronal controls themselves constitute a clock, or that they observe an accurate clock, whereby they trigger activation of the dreaming state at regular intervals throughout sleep. This is an automatic, intrinsic process. It looks, therefore, as though this activation of the brain for dreaming purposes must be very important in some way. We need to understand this better.

What Is the Purpose of Dreaming?

This is one of the great mysteries of the science of sleep at present. What is the function of the recurrent brain activation of dreaming during sleep? It is so prominent, almost dominant, in early life, that it attracts Western scientists to the idea that it may be important for the development of the brain. In later childhood and throughout adult life, although it is no longer so prominent, it nevertheless occupies a regular and a conspicuous part of one's adult life, 10 percent or more. You go from an intrauterine experience of 90 percent dream time to adult experience of the reciprocal of that. We might suppose that dreaming is perhaps necessary for maintenance of the brain in some important respect. This is what we would like to study for the next ten years: Exactly what benefits does this forebrain activation, this dreaming state, confer? We suspect that it has something to do with the capacity to maintain attention during the waking state.

ROBERT LIVINGSTON: It might be that dreaming has also to do with problem solving. Among other dream scenarios, the subject matter of dreams could well include, directly or by analogy, aspects of frustrated, unfulfilled, perhaps failed experiences.

There is abundant evidence from among scientists, artists, and others engaged in creative activities, that many of them have picked up cogent novel ideas and new patterns of thinking and performance during a night's sleep, and expressly during dreaming. Many of their most important challenges and frustrations have been cast in new light, opened to new strategies, or reformulated in entirely new and unexpected patterns for thought, action, and pursuit. These intellectually revolutionary solutions have emerged in consciousness during or promptly following dreams. Perhaps the dreaming state can try out alternative perceptions, judgments, and behavior without penalty of consequences.

ALLAN HOBSON: As yet we have no specific evidence bearing on that issue. We do not know whether cats dream, but they show brain waves indicative of activation associated with eye movements, and these occur periodically, at regular intervals, throughout their sleeping behavior. If we measure the electrical patterns of a cat's brain when the cat goes from nondreaming sleep to REM sleep, we see that the high-voltage slow electrical waves give way to low-voltage fast electrical activity. At the same time the skeletal muscle tone is actively inhibited and the animal is totally relaxed. So the brain is turned on and motor output is turned off except for eye movements.

Cells localized to the brain stem, in the vicinity of the nuclei that control eye movements, send signals to the visual receiving areas of cortex that process visual data. The electrical activity occurs just prior to the execution of eye movements. This means that the brain has a way of keeping track of movements even before the movements occur. Signals recorded in the visual cortex faithfully encode the direction and distance of the eye movements, which are actually taking place in the dark. The brain stem is telling the visual perceptual system where and by how much the eyes are going to be displaced, and hence what comparable shifts must occur between subject and visual field during whatever scenario is playing in the dream. The brain thus has a highly specific information system operating throughout dreaming. This probably is the physiological basis for the dynamic visual experiences associated with dreaming. Dreams are richly visual because the visual system is stimulated along with activation of the eye muscles.

We now have the interesting paradox that our brain, in the dark, with eyes closed, can turn itself on and initiate messages that relate to vision and other aspects of our life experience. This is, indeed, what a dream is.

DALAI LAMA: It is impossible, isn't it, for a person who has been blind from birth ever to see color in a dream state?

ALLAN HOBSON: It is impossible for them to realize this activation as visual because they have no categories of visual experience to bring to this synthetic task. This suggests that our visual system is entrained from exposure to visual events and that it is not simply intrinsic to that region of the brain. Blind patients must have had some previous visual experience in order to have visual imagery accompany their dreams. People with acquired blindness see in their dreams, and see only in their dreams.

Yesterday, Dr. Squire distinguished between declarative memory and procedural, or nondeclarative, memory. We might expect the forebrain, including the hippocampus and cortex, to be responsible for declarative sorts of memories, and other parts of the brain to be concerned with procedural memories associated with learned skilled behaviors. This part of the brain is constantly active throughout sleep and dramatically activated during REM sleep. In association with each eye movement, the activity of this part of the brain is altered, leading us to believe that one of the functions of REM sleep may be actually to rehearse brain programs for the benefit of behavior. This would provide procedural learning during dreaming. So the maintenance function theory of dreams now attains further specificity. What we may be doing every night in our dreams is rehearsing our basic motor skills, practicing to make perfect, if you like, making certain that the central programs for behavior are in good order. Although this is a speculative theory, it is provocative.

Lucid Dreaming

Dr. Hobson introduced the concept of lucid dreaming with an illustration of a man on a flying carpet. Flying is a favorite dream activity, and its obvious unreality provides a way of identifying the dream state.

ALLAN HOBSON: This man on a magic carpet, flying through the air, has taught himself to do this by taking advantage of some of the facts about dreaming. He knows that dreams are strange, that they have curious characteristics. So he tells himself before he goes to sleep that if he has a conscious experience that is strange, he will know that he is dreaming. He needs to do this for about three weeks, every night, for just a few seconds before going to sleep, and then he starts to gain consciousness of his dreams. He has created a new state in which part of his brain is acting as if it were awake while the other parts are dreaming. Western practitioners of this skill call themselves lucid dreamers. They can watch their dreams while their dreams are occurring.

DALAI LAMA: Is there special training that leads to that ability?

ALLAN HOBSON: It is simple. For three weeks you tell yourself before sleep that you ordinarily dream for about two hours, and that if you have a strange experience, you are going to recognize consciously that it is a dream. It helps to have a notebook at your bedside and to write down your dream experiences. You can induce lucid dreams by this automatic procedure, a split consciousness. In other words, the power of the mind is quite appreciable, capable of changing the state of the brain.

ANTONIO DAMASIO: Have lucid dreamers been studied over a long period of time?

ALLAN HOBSON: Lucid dreamers have not been studied over a long period of time in sleep labs. I suppose your question would be: What is the functional gain?

ANTONIO DAMASIO: What is the functional loss?

ALLAN HOBSON: Who knows? It could be that you are interrupting sleep in a deleterious way. All I can tell you is that their period of lucidity is fleeting. Even when they become highly skilled, the dream plot tends to slip away from them.

Until two years ago, there was no physiological evidence for lucid dreaming, and now there is. Subjects can be trained to make a prearranged sequence of eye movements so as to indicate that they are presently lucid, and these eye movements can of course be recorded electrophysiologically.

During lucid dreaming, general body muscle tone is still inhibited. But of course they have eye motor control which is presumably under the control of the frontal eye fields in cortex of the frontal lobes. As a means of testing and documenting lucid dream states, the subjects are told to make three full-excursion movements to the left followed by three full-excursion movements to the right. The probability of this occurring spontaneously is essentially zero

DALAI LAMA: In the Buddhist practice of dream yoga, there is a particular practice in which the lucid dreamer, while in the dream state, is told to be conscious of the dream state and told to meditate on something specific.

ALLAN HOBSON: Yes, it could very well be that this would work, and it could be experimentally documented, but it has not yet been tried in the Western scientific tradition. It's a doable experiment.

Now this finding has lots of important implications, and I would like to emphasize two. For Western science, it gives us a way of making time labels within the dream so that we can correlate physiological activity with mental activity more precisely. And it raises certain problems. It shows how suggestible dream contents can be.

What that means is that we can teach subjects to dream anything they want to dream about. Therefore, if the dream is taken as important evidence for a psychological or philosophical theory, we encounter the problem of a circular loop. The subject may be dreaming what he expects to dream about in order to prove the theory, and this does not constitute scientific evidence of anything.

DALAI LAMA: Are you implying that in psychology, where a dream may be considered of great importance for interpreting some of their theories, these experiments reveal that the psychological basis could be shaky?

ALLAN HOBSON: Yes, precisely. So that's the bad news. The good news is that this lucid dreaming state is so plastic that it can be exploited for a variety of purposes.

DALAI LAMA: From a Buddhist point of view, one might be able to distinguish different states of dreaming. Generally speaking, a dream is a dream, something you can't control. But for the highly advanced meditator, there could be possibilities for gaining certain insights through dreams.

ALLAN HOBSON: That's possible. Even at that level you encounter the same problem of wondering whether this is evidence of anything other than purpose and expectation.

DALAI LAMA: I know some Tibetans who lived in Tibet prior to the 1959 uprising. Before their escape from Tibet, they did not know about the natural trails and passes by which to get over the Himalayas into India. Some of these people I met had very clear dreams of these tracks and, years later, when they actually had to follow the actual trails, they found that they were already familiar with them because of the very clear dreams they had had previously.

ALLAN HOBSON: This is a so-called precognitive dream, and there are many examples of this in the West as well. I would like to defer discussion of that until later, as it is an important question.

Dr. Hobson closed his presentation with another drawing based on a dream image: a man riding one of two bicycles which are connected by an elaborate magical device.

ALLAN HOBSON: The good news is that dreaming is an autocreative state. Its plasticity can be utilized for a number of different purposes. This picture shows another drawing of a dream which indicates the autocreative nature of dream experiences. What does it mean that there is no one sitting on the second bicycle? The dreamer is a bachelor who throughout his dream journal complains of the fact that he has no companion. Now this is an interpretation and probably can never be proved scientifically, but it is nonetheless poetically intriguing. Dreams are often poetically compelling and we should not lose sight of that important point.

9. Manifestations of Subtle Consciousness

The discussion now turns back to the question of subtle consciousness. The tantric systems of physiology that are important in Tibetan Buddhism recognize manifestations of subtle consciousness in deep, nondreaming sleep, and in other states including orgasm. These are seen as opportunities to familiarize oneself with subtle consciousness in preparation for its manifestation at death.

The long-standing confusion surrounding the term subtle consciousness *begins to clear in this discussion. Tibetan Buddhism understands a broader range of meaning in the term* consciousness *than Western science, and* subtle consciousness *in particular may include mental activity that occurs in the realm that Western science classifies as* subconscious *or even during states which we normally consider unconscious.*

DALAI LAMA: In a certain body of Buddhist treatises known as tantra, there is one called the *Kālacakra Tantra*. A theory of this system identifies the following four states: the waking state, the dreaming sleep state, the nondreaming sleep state, and, finally, the orgasm. There is a kind of physical substance in the body, called "drops," associated with each of these four states, and each of those four kinds of substances has its specific source and location within the body.

You have established the relationship of brain states and functions with certain mental states, specifically the waking state, the dreaming sleep state, and the nondreaming sleep state. But, so far, we have not discussed the mind or brain state relating to orgasm.

In these same tantric treatises there is a lot of emphasis on the state of mind at the point of orgasm. Now there has been some discussion here concerning whether or not subtle states of mind exist. In Buddhism, the subtle state of consciousness is often not manifest, not evident. However, it is posited in these tantric treatises that at the point of orgasm, the subtle awareness then becomes evident. In fact, there are four occasions in which to varying degrees the subtle form of awareness manifests itself: orgasm, yawning, sneezing, and deep, dreamless sleep. In each of these four cases, to varying degrees, the subtle consciousness does become evident. Somehow the consciousness deepens.

ALLAN HOBSON: This is interesting. I would say that both orgasm and sneezing are state-dependent behaviors, and they are subsumed in our way of thinking among the other states. For example, sneezing only occurs in the waking state, and never occurs in sleeping. Orgasm occurs in the waking state, and it can occur in dreaming, but it does not occur in nondreaming sleep. Yawning occurs in the waking state, when one is sleepy.

DALAI LAMA: Do you see any physiological differences in the state of the brain during those four different states?

ALLAN HOBSON: The only work that has been done is in relation to yawning. There is some evidence of changes in the brain associated with yawning. Not very much has been done on orgasm in any state. The incidence of orgasm in dreaming sleep is so low that it is not easy to study. But lucid dreamers are very good subjects for this study because they can induce orgasm at will. It is one of the reasons people enjoy lucid dreaming, because they can have sexual pleasure without a partner, without social consequences.

ANTONIO DAMASIO: There are changes that happen during yawning and sneezing, but brain changes associated with orgasm are much more marked. They have to do with changes in oxygen delivery to the brain associated with changes in breathing patterns, changes in heart rate, sweating, vasodilation, and tremendously complex changes in chemical mediators, including dopamine, serotonin, acetylcholine, and endorphins. There is a tremendous change in a variety of chemical substances that directly affect the actions of neurons in different circuits. So you really can alter consciousness in the broad meaning of the term. Yawning and sneezing occupy much briefer periods.

DALAI LAMA: Do you find a common denominator in these states?

ANTONIO DAMASIO: It would be difficult to give an adequate answer. I believe that the knowledge needed to answer that question is too limited. Certain things come to mind: changes in oxygen delivery and distribution within the brain and vast chemical changes. So at least those two global shifts which can influence brain functions occur in each of the four states.

LEWIS JUDD: There isn't fully substantial data on these problems. And sneezing, at least, is of quite short duration.

ALLAN HOBSON: That is an important point. States relating to sleep might be said to be tonic in that they endure for long periods of time. In contrast, the first three examples you gave—sneezing, yawning, and orgasm—are phasic events which occur relatively rapidly and are over fairly soon. The neural mechanisms have to reflect that difference.

I should like to understand what, according to the Buddhist tradition, is the state associated with nondreaming sleep? How is it experienced? What are its characteristics?

DALAI LAMA: Within the Buddhist tradition, we don't speak in terms of the brain but rather of subjective awareness and also energies as these are experienced subjectively. Within that context, a distinction is made between grosser and subtler states of consciousness associated with grosser and subtler states of energy within the body. In deep sleep, the five sensory modalities have become inactive, and correspondingly the centers associated with them have become inactive. These changes are considered relatively gross. They also take place in a sequential process of going into deep, dreamless sleep, with these grosser states of awareness going dormant and the more subtle state of purely mental awareness becoming evident.

In the mind that is untrained in meditative practice, this sequence of the mind becoming more subtle will frequently not be evident. There are eight stages in this process of going into deep sleep. For a mind that is very finely disciplined in meditation, each of those stages will become evident experientially. In relation to the nondreaming sleep state, the dreaming state is understood to be somewhat more gross. And according to certain texts, there are physiological processes that correspond to these different mental states, and these are associated with subjectively experienced energies in the body.

To explain this more elaborately, we need to go into the whole system of channels and energy centers in the body, the *chakras*. But, without going into that for the time being, it can be said that in the waking state, these energies tend to be drawn into a locus here in the center of the head, at the level of the forehead. In the dreaming stage, these energies will be even more drawn to a point in the throat. In the deep sleep state, these energies are more drawn into the heart. The location is not the physical heart, the organ, but the heart center which is right in the center of the chest.

Certain events are experienced in meditation that seem to corroborate this theory. For example, in meditation, it is possible to bring your awareness into the heart *chakra*, and sometimes when this happens, the person will faint. On other occasions, the meditative awareness, finely concentrated, may be brought into the area of the navel. And at this juncture, it has been found experientially that heat is produced by such concentration. If you look at the anatomy of the body, you don't find these *chakra* points.

ALLAN HOBSON: I would like to come back to the question of the meditation associated with the eight levels of sleep in a trained subject. Tell us, for example, what are the characteristics of the deepest state so we can respond from the point of view of Western science.

DALAI LAMA: I don't know! Through training, at that deep level, awareness eventually becomes deeper, deeper, deeper. Then, finally, at the deepest experience, breathing stops.

ALLAN HOBSON: In general, I think that our experience is the same. But we must be very careful about the terminology. We distinguish objectively five stages in our formulation: waking and four stages of sleep.

DALAI LAMA: From a Buddhist point of view, the eight stages are associated with the dissolution of the five elements within the body. The five elements are the elements of earth, water, fire, air, and consciousness. Bear in mind, "earth" doesn't mean dirt, it refers to the solid constituents of the body; water refers to the fluid constituents, and so on. The dissolution of these elements takes place through five sequential states, and there are three more, corresponding to more and more subtle states of consciousness. The eighth is the most subtle.

ALLAN HOBSON: It might interest you to know that in nontrained subjects, when we perform awakenings in the various stages of sleep, we get less and less evidence of conscious experience. In other words, the deepest stage of nondreaming sleep, defined physiologically, is associated in our untrained subjects with very low levels of consciousness. And in 50 percent of the awakenings there is no evidence of consciousness at all. That doesn't mean that it couldn't be changed by training the subjects. But the evidence, from the point of view of the brain, is that you would need to provide a lot of training relating to the physiology.

THUPTEN JINPA: I think there is a certain misunderstanding of this term *consciousness*. In the Western philosophical or psychological term, consciousness is conscious, whereas, when Buddhist translators use the term, it has a wider meaning. It includes both the subconscious and unconscious levels as well.

ALAN WALLACE: The terms *conscious* and *unconscious* are not used in Buddhism. Rather, one speaks of differing degrees of clarity and subtlety of awareness. Even when someone has fainted, for example, and is regarded, in Western terms, as unconscious, from a Buddhist perspective a subtle level of consciousness is still present.

DALAI LAMA: As early as the seventh stage approaching deep sleep, you would say that your awareness has declined. That is, the mind is not clearly apprehending anything.

In Buddhist psychology, we refer to sleep as one of many mental factors. There are fifty-one mental factors, among which sleep is one. But it is said that the mental factor of sleep precedes the deep sleep state, meaning the nondreaming state, just as fear may proceed fainting without going into it. Consider an analogy: Once you have fainted, you feel no fear. However, it can be fear that leads to your fainting. So the fear doesn't go with you into the fainting. Thus, analogously, in the nondreaming deep sleep state, the mental factor of sleep has already passed.

It is a matter of terminology. Just how this mental factor of sleep in Buddhist terminology corresponds to Western terminology remains to be seen.

ANTONIO DAMASIO: I have a question, for clarification. When you have the very subtle consciousness that one is supposed to have just before death, that really does not mean that you have heightened, greater awareness, but rather the contrary. It means that in fact you are reducing perception to very low levels, right?

DALAI LAMA: Yes.

ALLAN HOBSON: That is very important.

PATRICIA CHURCHLAND: I had assumed exactly the reverse.

ANTONIO DAMASIO: I also had the opposite idea.

DALAI LAMA: One of the purposes of tantric meditation is to prepare you to be able to utilize the death opportunity. The point is to transform that stage of mind into wisdom, because it is the most subtle state of the mind. There is less influence of conditioning, so it is more pure.

ANTONIO DAMASIO: Mental exercise in preparation for death.

DALAI LAMA: Yes, it is very strong.

LARRY SQUIRE: Certainly, modern neuroscience is very sympathetic to the importance of the unconscious. The specialization and differentiation of the brain tells us that things are going on sometimes automatically, or sometimes in ways that we just don't have access to, given that we are using language mostly for our understanding as well as communication.

This is dramatically brought out in experiments with patients who have had the two halves of their brains separated surgically. In those people you can show that the right side of the brain can be getting information and doing things that the left side of the brain, which is the speaking half of the brain, does not understand. The left side may then make up things, confabulate, to explain behavior for which it cannot see the origins.

10. What Constitutes Scientific Evidence?

A persistent theme throughout the conference is the criteria for proof of an argument. How do we know what we know? When is it reasonable to generalize from particular observations? How do we deal with exceptional cases? Should testimony—allowable evidence in law—be completely discounted by science? How can the scientific method, which relies entirely on "objective" observation, begin to account for the subjective experience of consciousness? Both science and Buddhism rely on methods that constantly test belief against empirical experience, but Buddhism allows subjective experience as valuable evidence in the study of consciousness.

DALAI LAMA: There are certain people who feel they have out-of-body experiences while dreaming.

ALLAN HOBSON: This has not been studied in the laboratory, but it is easy to imagine how such a state could arise since it is possible to hallucinate practically anything during dreaming.

DALAI LAMA: There are accounts of people experiencing this sense of leaving their body, actually perceiving things in the external world, and later being able to recall events that presumably took place there, even to the point of being able to read a book in someone else's house. Has there been no scientific investigation of this type of testimony?

ALLAN HOBSON: That is correct, there has been no scientific investigation of these. But I would like to discuss this issue because I think that the issue of precognitive dreams, out-of-body experiences, and claims of previous lives, all have a problem in common for science. And in discussing this issue, I want to make clear, first, that my mind is not closed. But I am a scientist. So the opening in my mind is probably quite narrow!

Now this narrow opening, which is guarded by skepticism, is a crucial part of our scientific mental discipline, I believe. It is as important to our understanding of the truth as is inspiration. It is not a wish to ignore the truth. It is a wish to critically test belief against experience. So in that spirit, I think that Western science has a lot to offer, as a tool, and not as a weapon.

DALAI LAMA: Yes, very good. Beautiful.

ALLAN HOBSON: The question is: How can we advance any of these claims above the status of what we would call testimony and anecdote?

DALAI LAMA: The best thing is to experiment on those people who make these claims.

ALLAN HOBSON: That is one way. But then the question becomes: What kind of experiment? But before we get to that, let me add one more point that I think is important. In a court of law, testimony is important as evidence.

DALAI LAMA: Is it not held in science, that if some event or experience is true for a normal person, it must be true as well for all other normal persons?

ALLAN HOBSON: That generalization is not one of the demands that one makes. It is true that there can be exceptional individuals. We know that not everyone is a lucid dreamer, for example.

PATRICIA CHURCHLAND: It would have to be that every similar person must have that capacity. But sometimes it would be hard to tell whether two or three persons are similar in the right respect. Yet it would have to be that everyone who was similar to that person would have that property or share that experience.

ANTONIO DAMASIO: Or at least we need to have one example that would be convincing. If we would find proof that a cat can fly, we would not need to see any other cat fly. If only one had verified flight capability, we would be able to refute the statement that all cats cannot fly and we could deny that that's a true statement. All you need is one solid example of something and you have proof.

ALLAN HOBSON: The fewer the examples, the stronger the evidence must be.

DALAI LAMA: Yes, that's right.

ALLAN HOBSON: Generalization will help you, even if the effect is weak, because generalization is robust and widely distributed. So if you are going to rely on individual cases, exceptional individuals, then the evidence must be particularly convincing to overcome skepticism—and the skepticism in our minds is very marked, because there have been so many claims in the West, as in the East, of this sort of experience, which, when put to critical test, do not convince us as constituting evidence. It constitutes testimony, but not evidence.

ROBERT LIVINGSTON: The history of science is largely an account of disabusing ourselves of mistaken speculative suppositions. Elan vital and phlogiston are but two examples, each widely adhered to for decades.

ALLAN HOBSON: There is another important point I would like to discuss concerning method. Dr. Judd pointed out to me yesterday that there is an important distinction between a retrospective experiment and a prospective experiment. Maybe, Lew, you would comment on that and illuminate the point.

LEWIS JUDD: I think that in the Western scientific tradition, the basis of the ultimate, so-called scientific truth, is really created when you start from a certain point, establish a set of experimental conditions, and then watch it unfold. This is in contrast to looking back from a point to find evidence by which to establish scientific truth. So, for example, in one of the issues that was raised yesterday, about the memory of these two little girls of their past lives, we were saying that there are various ways one might explain that away. Not necessarily to discount it, but ways that would survive skepticism.

On the other hand, if one were to conduct a prospective experiment, starting from scratch, as to whether the information we are gathering today might play out in future lives, there might be a way to test future Dalai Lamas as to whether or not they remember neuroscience information that they acquired on this day, ten years from now, or fifty years from now, or five hundred years from now.

ANTONIO DAMASIO: May I add that any observations made up to now can be the basis for constructing a hypothesis, rather than being immediately taken as evidence. With that hypothesis, with that theory, then we can proceed to set up experiments deliberately to test whether the evidence supports the hypothesis and see just how strong the evidence is. In experimental research, we tend to go that way. An observation, an idea, leads to a hypothesis, a theory is constructed, and then we conduct experiments and decide how strong the hypothesis really is.

It tends to be the case that we never prove anything. In fact, all that we can do, very modestly, is to determine whether or not our experimental results are strongly in favor, or slightly in favor, or definitely against the hypothesis. We decide on the weight of all pertinent evidence.

LEWIS JUDD: Is Your Holiness familiar with the concept of prospective design of experiments? For example, let's say we have a hypothesis, a theory, that if a certain drug A is given to an individual it will suppress certain symptoms. That's the hypothesis. One would then design an experiment to give the drug or medication under controlled circumstances and measure the effects to establish whether the drug had the anticipated or other effects. To make the evidence more foolproof, you could conduct double-blind experiments where persons administering the drug and those measuring the effects would not know whether the individual were actually receiving that drug or some substitute drug or a placebo. There are a number of traditional ways that can be used to avoid subjective impressions creating biases and introducing noise into the evidence.

DALAI LAMA: This issue is very clear.

ALLAN HOBSON: Again, I wish to emphasize the point about creative use of skepticism. What Western scientists do, that to you might appear to be negative, is really their attempt not to be fooled. In other words, what a Western scientist does in constructing prospective experiments is to try to set the conditions in such a way that all alternative explanations will be eliminated and that his hypothesis can be disproved. Thus, an important spirit of science is to be open to giving up the hypothesis. That's an important attitude, and difficult to fulfill.

DALAI LAMA: I suspect that sometimes scientists, too, tenaciously cling to a hypothesis so much that they still adhere to it, regardless of contrary evidence.

LEWIS JUDD: Yes.

ALLAN HOBSON: Skeptical investigation is very important.

ANTONIO DAMASIO: Actually, one measure of the quality of a scientist may be how readily he gives up his cherished hypotheses. In fact, we are consistently adhering to very good ideas which may in a decade or two be clearly shown to be wrong. And there are some people who simply can't give them up and will cling to them.

DALAI LAMA: On one occasion, I met a group of scientists who introduced themselves to me one after the other, and one of them told me, "I am a fanatical materialist. I will not accept the existence of mind." So my question is, for these extreme, radical materialists who refuse mind's existence, insisting that it simply does not exist, why are they saying that? What do they mean by the term *mind*?

PATRICIA CHURCHLAND: What they mean by mind is the brain. That's what they mean.

DALAI LAMA: Nobody can deny the existence of the brain.

PATRICIA CHURCHLAND: When they said the mind doesn't exist, they mean there isn't something spirit-like that is independent of the brain. They think of perceptions or thoughts or dreams as processes of the brain. And that was really what I was talking about yesterday. There is a common idea that there is a nonphysical soul, but when you look more closely at what neuroscience has discovered it looks like there is only the brain.

DALAI LAMA: Even in Buddhism there is no notion of a self-sufficient, self-supporting "I," self, soul, or ego. This is thoroughly refuted. Buddhists do assert the existence of awareness, but to use the term *soul* in a Buddhist context is misleading, because Buddhists don't use the term and by and large refute the existence of a soul.

PATRICIA CHURCHLAND: But do you think that there is something, I am not sure what to call it—a kind of awareness that can exist independently of the brain? For example, something that survives death?

DALAI LAMA: Generally speaking, awareness, in the sense of our familiar, day-to-day mental processes, does not exist apart from or independent of the brain, according to the Buddhist view. But Buddhism holds that the cause of this awareness is to be found in a preceding continuum of awareness, and that is why one speaks of a stream of awareness from one life to another. Whence does this awareness arise initially? It must arise fundamentally not from a physical base but from a preceding continuum of awareness.

LEWIS JUDD: So it is independent of brain function.

DALAI LAMA: The continuum of awareness that conjoins with the fetus does not depend upon the brain. There are some documented cases of advanced practitioners whose bodies, after death, escape what happens to everyone else and do not decompose for some time—for two or three weeks or even longer. The awareness that finally leaves their body is a primordial awareness that is not dependent upon the body. There have been many accounts in the past of advanced practitioners remaining in meditation in this subtle state of consciousness when they died, and decomposition of their body was postponed although the body remained at room temperature.

LEWIS JUDD: How would you know that their brain function had completely ceased?

DALAI LAMA: This would be good to check out with the instruments of Western science. Last year there were two advanced practitioners, one of whom remained in that state without his body decomposing for four days, another for about ten days. But unfortunately, on those two occasions there was no physician on hand to conduct the test. We have to make arrangements before such things happen. Hopefully, we will have equipment and an expert available to fulfill this objective.

ALLAN HOBSON: There again, you see, skepticism would immediately try to imagine other explanations for the same phenomena, namely that the brain might still be active at a low level. We know that there can still be valid neuronal activity even when the EEG is absolutely flat. There still may be neurons down in the brain stem which are firing. This happens, for example, in hibernation.

ROBERT LIVINGSTON: This is a difficult experiment. Western science has dealt with brain death for quite a while. Brain death is very tricky to establish. Whether brain death can be established beyond peradventure of doubt may be crucial as to whether you continue to provide life support for that individual, or not. I think that if you had a person dying in that way, it would be very difficult to decide when you might have reached some level of "subtle awareness."

DALAI LAMA: The methods for realizing this state of awareness are set forth in Buddhist treatises on philosophy and meditation. We follow the methods and we see outward signs, but we can't tell for certain what they indicate. What we see externally is that the meditator, who is well into the dying process, is sitting upright in meditation, with no physical movement, heartbeat, or respiration. And he stays like that for four to ten days.

Prior to the cessation of breathing and heartbeat, the body is already degenerating because of illness. But now, following the cessation of breathing and heartbeat, the metabolism, as far as we can tell on a gross level, becomes restored somewhat. These interpretations are based on external observations.

ALLAN HOBSON: And they are signs which imply the continuing existence of the body. We would say skeptically that the brain is probably still alive. You would say the practitioner's awareness persists in some subtle but not nonphysical way.

DALAI LAMA: Yes. You are saying that some part of the brain stem is still functioning?

ALLAN HOBSON: That is an alternative explanation.

LEWIS JUDD: It is a hypothesis that needs experimental testing.

DALAI LAMA: Good, let's do it!

DALAI LAMA: In Buddhism we speak of three types of phenomena: First, there are evident phenomena that are perceived directly.

Second, there are slightly hidden phenomena, which are not accessible to immediate perception. There are differences of opinion on this even within Buddhist philosophy. Generally speaking, we think this second type of phenomena can be known indirectly by inference. One example of something known by inference is that anything arising in dependence upon causes and conditions is itself subject to disintegration and momentary change. This momentary change is not immediately evident to your senses. You can look at something with your eyes, and it does not appear to be changing right now, but by inference you can know that it is momentarily changing. This is an example of the second category of phenomena.

Third, there are very concealed phenomena, which cannot be known by either of the two preceding methods. They can be known only by relying upon testimony of someone such as the Buddha.

Leaving aside the third category, do you as scientists accept the first two categories?

PATRICIA CHURCHLAND: The first two categories seem to be roughly the way Western scientists also think about them.

ALLAN HOBSON: The more we know about phenomena, the less we need the third category. My contention would be that your knowledge about the first two categories, and especially the second, is limited to the degree that we impute the third category. As our knowledge of the second category grows, our need for the third category will diminish.

DALAI LAMA: The same could be said about the second category as well. If we can increase our ability to see things perceptually, then the second category diminishes.

ALLAN HOBSON: Absolutely. That's the task of science.

DALAI LAMA: This threefold categorization—in the Buddhist context—is not coming from some inherent differentiation among phenomena, but rather from the limitations of our capacity for awareness.

For example, something that may be slightly concealed for me may be evident for another person. What is occurring in my mind right now is to me evident, but for you it is concealed. Unless I tell you how I feel at the moment, there is no way for you to know. Apart from that testimony, there is no access to it.

PATRICIA CHURCHLAND: Yes, except that we can rely on other aspects of your behavior. That is, if you were wincing or holding your jaw, I would infer that you had a toothache even if you didn't say so. Thus, there would be other aspects of your behavior, body language, and so on, that would give us information. Your testimony is of course important, but it is not decisive. You could be playacting.

ROBERT LIVINGSTON: It is the problem of a secret. You can hold a secret, and nobody can discover that secret.

DALAI LAMA: Here is my point: What do we have to say right now about those phenomena that can be known only by testimony, such as what I am thinking right now?

LARRY SQUIRE: That is a practical statement. Neuroscience would say that, in principle, with enough technology, these things would become accessible—for example, if there were techniques by which to measure fine structural activity in many parts of the brain directly and simultaneously.

ALLAN HOBSON: Let me give you one very strong example. I could tell that you were dreaming even when you didn't know you were dreaming.

DALAI LAMA: Can you tell what I am dreaming?

ALLAN HOBSON: Wait a minute, wait a minute, be patient!

ANTONIO DAMASIO: One could describe the agenda of science—what we want to do, and in a very modest way what we are already doing—is to reduce our reliance on that third category. It is obvious that we are not going to complete this for a very long time, but what we want to do is to remove more and more issues from the third category and move them into the first two categories.

LEWIS JUDD: And from the category two into one. We are always pushing issues into category one. That's the importance of science.

ANTONIO DAMASIO: The process is always shifting, based on better observations, better technology, and better theory.

DALAI LAMA: I am speculating that perhaps as a result of increases in scientific understanding and technical advances, things which might have been extremely concealed, as Buddhists talk about these categories, might even become clearly evident; for instance, like the earth being round. Centuries ago, if you believed that, it would be only on the basis of testimony, because somebody said so. Now you can see it with your own eyes, in pictures.

ROBERT LIVINGSTON: I can give you a concrete example of the progress of our learning about how consciousness and personality depend on detailed brain structures.

We studied scores of postmortem human brains, from people who had experienced no known neurological or psychiatric illnesses. By slicing whole brains at microscopically thin intervals and imaging each freshly cut surface from a fixed camera position, we obtained motion pictures in perfect registration of each successive surface all the way through each brain.

Examining these images, we found that the surface contours and internal structures are vastly different from each brain to any other. The differences we found are grossly obvious. There are twofold differences in areas of cortex that have important, discrete functions, and similar gross differences in subcortical regions. We found every brain to be unique, just as our faces are. But the differences in brain structures are deeply meaningful in respect to perceptions, memory, motor and emotional skills, judgment, personality, and character.

We made computer reconstructions of one brain so that the principal structures could be seen three-dimensionally and dynamically rotated and moved about for purposes of demonstration in a documentary film. We believe that computers and suitably expanded memory management systems will allow detailed quantitative comparisons to be made of differences in microscopic detail among many whole human brains.

Comparisons between detailed life histories and the detailed structures of the corresponding brains will allow what our friends call "endophrenology." This technique would permit correlations between details of brain organization and human qualities of consciousness, perception, judgment, temperament, and behavior. Of course, such data can also be used to compare magnetic resonance images of living brains, and thereby obtain even more fascinating subjective correlations.

At some point we should be able to say that certain notably expanded structures in a particular brain suggest that this individual likely had great musical ability, and so forth. Perhaps by observing brain structures that are especially noteworthy in certain remarkable individuals, we shall be able to identify brain morphology that is characteristic of compassion. I confidently predict a branch of neurosciences that will disclose features of the human brain, presently largely concealed, that relate to both our inner subjective and our outer worldly life experiences.

DALAI LAMA: Although it is difficult to pinpoint the physical base or location of awareness, it is perhaps the most precious thing concealed within our brains. And it is something that the individual alone can feel and experience. Each of us cherishes it highly, yet it is private.

ALLAN HOBSON: But then, when we share, if we dare, what is going on, it usually tends to reveal many things in common. So what appears to us to be inevitably private is in fact quite generally shared.

ANTONIO DAMASIO: It is much more than that. It has tremendous commonality. It is interesting to think about differences and similarities in both brains and minds; it depends on your perspective. On the one hand, there are differences in every brain. Each is unique. At the same time, they are generally similar. The same with our spirits, we are remarkably similar in spirit. That is how you can have human sympathy, empathy, and cross-cultural compassion.

11. Psychiatric Illnesses and Psychopharmacology

Lewis L. Judd, MD

Lewis Judd is a psychiatrist whose research has focused on psychopharmacology: the effects of drugs on the brain. At the time of this conference he was the director of the National Institute for Mental Health (NIMH), the primary federal agency concerned with the study of mental illnesses and providing leadership for the treatment of the mentally ill.

He presents an overview here of recent developments in his field: the growing understanding of mental illnesses as biomedical disorders, the importance of noninvasive imaging techniques and pharmacology to this understanding, and the impact of a systematic classification of mental illnesses for diagnostic purposes and treatment.

LEWIS JUDD: What I would like to do today is to give you a feeling for how our concept of mental illness has developed, in this country and in the West more generally, a concept which has been undergoing considerable change during the last twenty-five years. I also want to give you a sense of our current ideas about this field. I shall be very interested to learn from your side what commonalities might be shared with Tibetan medicine, which is so richly interwoven with Tibetan Buddhism.

The reason that we in the West have been undergoing a considerable shift in our understanding of mental illness is because of some of the very things that you have been exposed to during the last day and a half. The growth in brain sciences has had an important impact on our understanding of mental illness. There has been an increasing trend within the field of mental illness to begin defining distinct entities of mental illness and their specific characteristics. In addition, as an outgrowth from some of the brain sciences, there has been a virtual explosion of information coming from the very important branch of pharmacology known as "psychopharmacology," which is the study of the effects of medications on brain and mental functions. Another important recent influence is the growing understanding of the role of genetics as a causal factor in mental disorders.

Currently we see mental disorders very much the way we view medical illnesses. That is, mental illnesses are essentially biomedical disorders that stem primarily from abnormal functions of the brain. We are now convinced that mental illness in its seemingly infinite variety is made up of a series of highly discrete disorders that have their own characteristics, their own symptoms and signs, such that they can be recognized and diagnosed as independent entities with specific treatments. Further, we now are convinced that a number of the most significant and severe mental disorders are controlled, at least in part, by genetic inheritance. I will touch on each of these topics briefly and present some evidence as to how we have come to this current concept of mental disorders.

The neurosciences are obviously a relatively new branch of science, coming into full fruition in the last thirty years. To give you an idea of how fast this area of science is growing, the Society for Neuroscience started in this country in the early 1970s with 200 scientists. There are now over 15,000, and we have another 15,000 in training, supported by funds from the National Institute for Mental Health and other institutes in the federal government. Neuroscientific research is a major enterprise in this country, and it is growing in other Western countries. In fact, it is about to receive a major boost in the United States because within the last two months the Congress, both the House of Representatives and the Senate, passed a Joint Resolution to declare 1990 as beginning the "Decade of the Brain." This is to launch a major national science effort to demystify through research what is certainly our most mysterious and complex organ system, our brain. This national endeavor will be the equivalent, we hope, of going to the moon, but in this context, we shall be examining inner space.

Growth of knowledge in the field of neuroscience has been phenomenal. Our staff at the institute did a survey within the last six months in which we found that 90 percent of what we now know about the brain has been published within the last ten years. So we are on an accelerating curve of discovery with respect to mind/brain problems. As new findings have emerged, they have had a powerful impact on our understanding of mental illnesses, and on our understanding of mental disorders. Right now we are seeing that brain dysfunctions usually lie at the core of mental disorders. Much as someone with cardiac failure has a diseased cardiovascular system, or someone with insulin-dependent diabetes has a diseased pancreas, people with mental disorders have dysfunctional brain structures.

Thus, we are conceptualizing mental illness in ways very similar to those described by Dr. Damasio when he pointed out to you yesterday how highly specific insults to the brain create very discrete deficits in the way people perceive, memorize, and think. Except that the brain lesions of the mentally ill are often not so discrete or delimited. They are more diffuse in nature, often spread throughout a wide variety of structures, and still very elusive. Yet we believe that we are beginning to find definite evidence of relevant disorders present in the brain.

Noninvasive Imaging: A Window on the Brain

One of the technical developments that has helped us enormously has been the new capacity within the last decade to get detailed pictures of the internal structures of the brain in living patients, without doing them the least harm. This gives us windows on the functioning brain and constitutes a major step forward. We had been held back in our understanding of mental disorders for centuries, because we could not readily get detailed information on brain function and structure that was sufficiently sophisticated and detailed to address the kinds of functional abnormalities we encounter with mental disorders.

Dr. Judd illustrated his point with a series of magnetic resonance image (MRI) scans. The first was from a study of identical twins then being conducted at the National Institute of Mental Health. The study compared the brains of pairs of identical twins, in which only one member of the pair was schizophrenic. He noted that, in terms of genetic inheritance, identical twins share about 85 percent of the same chromosomal material. The scan showed that in the twin afflicted with schizophrenia there was some loss of neuronal mass which had been filled in with cerebrospinal fluid from the ventricles.

LEWIS JUDD: Another MRI scan relates to a very severe disorder of children called autism. As a clinical syndrome, autism has been recognized since the 1940s. Originally it was attributed to cold, unemotional parenting. We have now discovered, using imaging techniques which have recently become available, that in autistic children there may be a lag in development of the cerebellum. This suggests that we are dealing not with a problem of poor parenting but with a severe developmental delay or arrest in this particular brain structure.

Autistic children have very late development of language, and sometimes they never develop language capabilities. They are often severely retarded intellectually. They do not relate well to the human social environment. The diagnosis of autism designates a child who is independent and aloof, not interacting with his human environment, not even with his parents and people the child knows well. An autistic child tends to be fascinated with things, with mechanical devices.

DALAI LAMA: Is it possible to recover from this structural defect, to create normal functions?

LEWIS JUDD: In autistic children it is not. There may be some later structural development of the cerebellum, but that is accompanied by minimal changes in the child's outlook and behavior. With medication, we can control certain behaviors of the child, such as self-destructiveness, but there is very little that can be done to rehabilitate the child in a comprehensive way.

I shall give you another example of how imaging techniques help us look at mental illnesses for which we previously had no evidence for abnormal brain structure or function. Using composite pictures of the brain obtained during dynamic changes in its function, we can show changes in blood flow from one part of the brain to another, indirect evidence of changes in biochemical or metabolic action. Nerve cells in the brain take in primarily the simple sugar glucose, oxygen, and, in lesser amounts, amino acids. In this particular study done at the NIMH, we are measuring the actual shift in blood flow from one part of the brain to another in a group of normal individuals compared with a group of schizophrenic patients.

Schizophrenia is a profound disorder of the brain that results in severe problems of thinking and cognition. Schizophrenic patients suffer from incoherent thinking. Individuals may have delusions, which are stable, false beliefs that they hold despite evidence to the contrary. They may also have hallucinations: they may hear voices, see things, or smell things that are not present. They often have a very difficult time adapting and living in the world. Frequently, they cannot take care of themselves. They may experience a profound absence of feelings and emotions, or their feelings and emotions may be quite inappropriate. Things they say may not be accompanied by appropriate expressions of feeling. They might, for example, describe something very horrible and sad, yet laugh as they do so. Their speech, behavior, mood, and feeling states may be quite disparate and incongruent. Schizophrenia is the most severe mental disorder that humans experience. It is most often a progressive disorder, the course slowly increasing downhill throughout the lifetime of the individual. They may end up being very, very disabled after a few years.

The brain scans show that while resting, the brains of normal subjects and schizophrenics look very much alike. But they differ considerably when we ask each individual to perform a specific task that requires an ability to abstract, an ability to solve problems, and an ability to remember. In the brains of normal individuals the prefrontal cortex "lights up," indicating that there has been a major influx of blood into that particular region. This shift of blood flow arises from the need to support increased neuronal activities required to carry out those challenging thinking processes.

In the schizophrenic patients, however, there is essentially no change in blood flow to that region. This quantitative information about regional blood flow allows us to focus on a specific region of the brain of schizophrenics that appears to be highly dysfunctional.

That is just an example. There are hundreds of studies of this nature that have established, we believe irrefutably, that the seat of the schizophrenic problem, and many other major mental disorders, resides in the brain.

Classifying Mental Illnesses

Another major step in clinical research that has been advanced in part by findings in the neurosciences has been the attempt to develop a detailed, accurate classification system, based upon empirical observations and accurate diagnostic criteria, to differentiate each of the many discrete expressions of mental disorders. This thick bookcontains detailed descriptions of mental disorders as conceptualized in the West, along with the various clinical characteristics, signs, and symptoms that one must identify in order to make a specific diagnosis of one type of mental disorder or another.

Mental disorders are discrete. In pure form, they do not closely resemble one another. They differ systematically and manifest their own identifying characteristics. For example, here is a description of major depression. Perhaps the second most common group of mental disorders is what we call "mood disorders," of which the most important one is major depression. Let us suppose that a clinician who is well trained sees someone who tells him or her, "I am profoundly sad, blue, and dysphoric. I have been so for several weeks. It doesn't change. I have problems in sleeping. I go to sleep, but I wake up every morning at two or three o'clock, and can't get back to sleep. I have lost my appetite. I have lost considerable weight, up to, say, 15 percent of my normal weight. I am unable to think and concentrate at the level that I used to. It is a problem every day. I have low energy. I can't accomplish anything. I am besieged with thoughts of death and dying and suicide."

The clinician knows almost immediately that, according to this accumulated cluster of symptoms, this individual suffers from major depression. This classification system helps him to confirm that diagnosis, and it tells him, in addition, once the diagnosis is made, what means he has by which to respond. This disorder, like so many other mental disorders, won't go away rapidly by itself. Major depression, if untreated, lasts about nine or ten months. If major depression is treated appropriately, the physician can usually bring relief rather quickly. So this diagnostic codification has made, we believe, a major contribution to our understanding and ability to respond to a spectrum of mental disorders.

Because of the accuracy of classification, and because we can thereby better train people to diagnose accurately and recognize various mental disorders, three years ago the NIMH conducted a major national survey of mental disorders existing in the U.S. population. We examined door-to-door a sample of adults in all age brackets, representative of both urban and rural areas. A total of 18,000 persons participated in specific research-structured diagnostic interviews. What we found was something that I think those of us who are clinicians already knew. Mental disorders, rather than being rare diseases, are very common, perhaps the most common category of diseases that mankind experiences. We found that between 12 and 13 percent of those interviewed had already experienced some recognizable mental disorder. They were, or had been, or should have been patients. So the scope of mental disorders in the United States has now been estimated quantitatively for the first time. Since this was a household survey, it defines a major public health problem.

DALAI LAMA: I am impressed by the numbers. The percentage seems quite high. Perhaps I should have been included in that survey!

LEWIS JUDD: One in ten, approximately. Moreover, we found that in terms of projecting lifetime risks, that percentage actually rises to 20 percent. That indicates that one in five persons, at least in the United States, will have a serious, diagnosable, and treatable mental disorder some time during their lifetime. These are serious, common mental disorders that are of major public health importance. We are beginning to move forward to address that problem full scale.

Advances in Psychopharmacology

In addition, paralleling this advance, there have been developments in the field of psychopharmacology, which is my area of research. It involves attempting to find medications that can help treat and improve mental conditions. For example, we now have available in the West more than thirty medications that are effective antidepressants, useful in treating people. This spectrum of medications is so broad, effective, and fundamentally sophisticated, that we can now manage to treat effectively 85 percent of all depressive disorders that gain the attention of trained clinicians.

DALAI LAMA: In the case of people who are depressed, are there not occasionally valid reasons why they are depressed, as when they reflect on certain tragedies and misfortunes, whether true or untrue? If as a result of their reflections or for some other reason, they get depressed, would this medication really help to reduce their depression?

LEWIS JUDD: First of all, antidepressants don't treat unhappiness. Antidepressants are highly specific to correct major depression, the syndrome I described that, in our view, is a clinically significant depressive disorder with a variety of well-defined characteristics.

DALAI LAMA: Isn't it possible that a mental affliction clinically identified as depression could arise from sustained thinking on some unfortunate circumstance? Doesn't that happen?

LEWIS JUDD: Absolutely, but it doesn't matter. We are looking at a continuum of biological vulnerabilities. We believe that virtually anybody can develop a major clinical depression. We know, or at least strongly suspect, that some people who develop depressive disorders inherit a genetic vulnerability to do so. In these susceptible cases, it doesn't take much provocation in the environment to trigger a major depression. There are others who don't have that vulnerability, who almost seem almost immune to depression. They may be exposed to all kinds of terrible things without becoming depressed, but still they can become depressed if there is an enormous accumulation of untoward, unprecedented, or tragic events in their lives.

What we are finding is that once you develop a clinical manifestation of a major depression, it doesn't matter much what has caused it. Once a person becomes depressed, then that condition needs to be treated specifically or it will persist for nine or ten months, or longer.

DALAI LAMA: Does major depression arise initially from some external cause that brings about a harmful change in the brain, leading to the symptoms, or, alternatively, is the original cause found in the brain and the illness is just triggered by something from the environment?

LEWIS JUDD: There probably is room in the model for both types. However, conceptually, I think we are looking at this as a genetic/

environmental interaction. Let us suppose that someone has a very, very high genetic propensity for developing depression. It may look as though the depression is being triggered from within, independent of what is happening in their environment. However, if you go back and look carefully at what happened to the person just before he or she began to show symptoms, you can always find some type of an "assault," perhaps subtle things, that may have triggered the depression. Because that person's threshold for developing depression is so low, it doesn't take much. So I would say, it takes two to tango—a genetic vulnerability and an environmental stressor. Major depression is a complex interaction between one's inherited constitutional givens and environmental events that elaborate and precipitate manifestations of the depressive disorder.

ROBERT LIVINGSTON: Lew, if I understood you correctly, you said you had something like thirty different kinds of medications that are helpful in coping with this syndrome of depression.

LEWIS JUDD: That's correct. There have really been almost three generations of antidepressants developed. Each of these generations has been more sophisticated and specific than the others. So, as I say, we can now manage about 85 percent of all depressive disorders. Certainly, in this country, you need not suffer long from a depressive disorder. We can absolutely manage this problem for the vast majority of cases.

LARRY SQUIRE: It might help to clarify to what extent Western science says that the drug is the treatment of choice, as opposed to improved understanding as the form of treatment.

LEWIS JUDD: First of all, I have been talking this far exclusively about medications that help specific disorders. But many medications are given to patients as a part of an interactive healing relationship between physician and patient. Any dispensing of medications has to be couched as part of a credible and caring transactional education about the illness from the patient to the physician and from the physician to the patient. Also, for some cases of depression, it is appropriate that we apply certain forms of psychotherapy that have been designed specifically to deal with individuals who have depressive disorders.

We are finding that if you give a patient with severe major depression psychotherapy only, it is often not very effective. But, psychotherapy can be very helpful for less severely ill patients with clinical depression. If you give medications only to severely depressed patients, it will help more than 60 percent of such patients. But, if you combine psychotherapy of a specific nature with the dispensing of appropriate medication, you raise that rate of relief even further. This is evidence that there may be a synergistic effect between psychopharmacology and specific forms of psychotherapy.

At present, we are convinced that this should not be an either/or therapeutic approach: either an appropriate medication or an appropriate kind of psychotherapeutic/educational transaction. Therapy often works best when these are combined. Certainly there is a tendency in this disorder for symptom recurrence. A high proportion of those who develop a depression will have at least one other depression. With patients who suffer repeated recurrences of depression, psychotherapy during the interval when symptoms are in abeyance will reduce the likelihood of further recurrence.

DALAI LAMA: Is continuing the medication by itself not helpful for lowering the repeat rate?

LEWIS JUDD: Medication maintained at the appropriate dose level is also very helpful. But, in addition, patients can be educated to avoid certain kinds of circumstances that are especially problematic for them or to manage such circumstances better so that they don't find themselves being stressed and psychologically assaulted in ways that may bring on depression.

DALAI LAMA: Of the two causes—physical disorder of the brain and exposure to distressing circumstances—which tends to be the initial cause? Is thinking about tragic circumstances the dominant cause?

LEWIS JUDD: Yes, it might be so. I think our experience indicates that depressive disorders can be created and sustained and potentially prevented by means of influential techniques that might be applied with respect to managing one's environment.

DALAI LAMA: Leaving aside the question of moral and ethical values, have there been cases where someone has had one brain image taken at the initial onset of depression and another at a later time without receiving medications? Has it been possible to observe brain changes while the depression is developing?

LEWIS JUDD: That has at this point not been done.

DALAI LAMA: You can do this with animals?

LEWIS JUDD: There are models for depression in animals that are experimentally manipulated. The best one that we use right now, which we think is analogous to depression in humans, is called "learned helplessness." You put the animal into a situation where basically it can't win, and then soon the animal gives up and looks very lethargic, as if it were depressed. We have studied the brain chemistry of animals in such states and found some highly specific things. However, it is not clear whether this is exactly equivalent to a depressive disorder in humans

The Genetic Inheritance of Mental Illness

Let me now present an example of a disorder that appears to be more regulated by inheritance than necessarily by environmental interaction. There is a very serious disease that has, according to virtually every study that's been done in other cultures and in other countries, a prevalence rate that is almost identical with what we are finding in the United States. Even though no other country has done the extensive epidemiological study that we have done nationally, the fact is that smaller studies elsewhere provide data that are very similar.

The disorder I'm talking about is manic-depressive illness, which attacks around 1 to 2 percent of the adult population. In the United States there are probably two and half million people with manic-depressive disorder. These are individuals for whom it runs in their family. There are very clear genetic factors in this disorder.

People with manic-depressive illness experience intense periods of depression, lasting anywhere from six to eight months, followed by a period of symptom quiescence, and then maybe a period of what we call manic behavior, which is the opposite of depression. This involves an intense agitated elation, euphoria, grandiosity, irritability, lack of need for sleep, boundless energy, and very poor social judgment.

These manic-depressive individuals have an average of five to eight episodes lasting from six to nine months each time throughout their lifetime. Sometimes they may have as many as fifty to sixty episodes in total. These episodes appear to attack people almost out of the blue. Once initiated, symptoms appear to be locked in, lasting for a roughly predictable period of time.

We have discovered a medication that is highly specific for this disorder: lithium, one of the trace elements found in the human body. Now, if you trace the natural course of the manic-depressive disorder in a patient before and after they have undertaken a course of lithium treatment, there is a phenomenal difference. Before we had lithium, manic-depressive patients spent approximately 25 percent of their adult lives in hospitals. They spent an additional 25 percent of their lives going into and coming out of those episodes. So it cost them roughly half their adult lives. Lithium was introduced as a medication in this country relatively recently, being approved by the Food and Drug Administration in 1969.

We have calculated that the introduction of that one drug alone, in this country has saved $39 billion in costs over the last twenty years: $12 billion in elimination of the need for hospitalization, and $27 billion in recovered productivity among individuals who were previously highly disordered and disabled and who are now living very productive lives. What we have discovered through studies on manic-depressive psychosis is a tight linkage—for the first time—between a highly specific diagnosis, a specific medication, and a predictable clinical benefit.

One of the most exciting areas of genetics has been discovered from population genetics and family pedigree studies, for example, from twin studies. This relates to both manic-depressive illness and to schizophrenia, both being mental disorders that run in families. If one identical twin has schizophrenia, the likelihood of the other twin having schizophrenia is about 35 to 45 percent. It is at least thirty-five times more likely to occur in the twin of a person who has schizophrenia than it would in the normal population, so there is a strong genetic preponderance.

We have been looking for families in which there is a genetic concentration of disorders. We are beginning to develop family pedigrees to track the offspring of parents with schizophrenia. In one family, for example, where one of the parents had schizophrenia, among their eleven offspring there were five with schizophrenia. Among their grandchildren, there were two. We are now on the lookout for families of this kind, to study their genetics, to analyze their chromosomes, and to find out whether we can identify the location of the chromosomes that are potentially responsible for at least a predisposition to these illnesses.

DALAI LAMA: In terms of hereditary diseases, if the illness is on the father's side, is there evidence that the boy or the girl from the family will be more likely to inherit that disease?

LEWIS JUDD: It depends on whether the inheritance is sex-linked. If it is linked to the sex or X chromosome, then it would reveal a preponderance effect on the son or daughter. For example, in major depression there is a two to one preponderance of women over men. In manic-depressive disorder, women are slightly more affected than men, but it is almost even. In schizophrenia, it is evenly divided between the sexes.

We know in manic-depressive illness that it is genetically transmitted. We strongly suspect that this is also true in schizophrenia. There may be other disorders that we are now finding that may be genetically linked, like obsessive-compulsive disorders and certain forms of anxiety disorder.

LEWIS JUDD: I have a question. The way we in the West presently think about mental disorders is fairly deterministic. For example, in an extreme case, with a high genetic tendency, in an environment filled with elaborate stresses, it is inevitable that someone will develop a mental disorder at some point. On this account, we are becoming increasingly less judgmental about the appearance of mental disorders and are disinclined to attribute them to past failures or to emotional weaknesses and various other factors.

What are the Buddhist conceptions of mental disorders? Is there attribution to some type of failure of self, or failure of enlightenment, or failure of centeredness, that would result in a mental disorder such as an increase of wind or whatever other humor might be involved?

DALAI LAMA: As I mentioned previously, from a Buddhist perspective we think of consciousness and energy as they are subjectively experienced. Within this context, then, if a person is experiencing some kind of mental dysfunction, it is frequently understood that the mind itself has become too withdrawn in upon itself and that there is a corresponding physiological process involving the energies themselves, which are closely associated with consciousness, also entering into a dysfunctional state.

So, in the Buddhist view, it can happen, for example, that one's mind will become depressed because of some environmental event. As a result of the mind becoming depressed, there is a chemical, maybe an electrochemical, transformation in the brain that has now occurred. The mental dysfunction will then be aggravated. When that happens, there is a further chemical response, which then avalanches upon itself. This is the Buddhist view, simply stated. It was with this in mind that I was asking previously, which has the greater dominance, external circumstances or internal ones?

Additionally, in the Buddhist view—and similarly in Western interpretations—on occasion, without any special external event taking place, there can simply be a dysfunction or disruption in the balance of the elements within the body. In that event, the internal circumstances are the dominant, principal cause. In dependence upon this physiological cause then, the mind can become depressed.

LEWIS JUDD: I was inquiring about something else as well. In Western society, having a mental disorder is still highly stigmatized. Someone is essentially at fault. They have presumably done something wrong. Because of this cultural attitude, oftentimes people are ashamed, and unable or unwilling to seek help. Is there a similar kind of pejorative social conception that affects the mentally ill in Tibet or India?

DALAI LAMA: It is more an individual matter. In a Buddhist culture, you have some who respond with compassion, and you have others who do not respond with compassion.

12. The Limits of Intervention

As our understanding of the relationship between the mind and the brain grows, technologies based on this knowledge have profound implications for mental health. But even as it grows, our understanding is limited to a small fraction of the brain's complexity. Is it possible from this limited viewpoint to judge the practical and ethical boundaries of future technologies?

DALAI LAMA: The preceding discussions seem to indicate that specific mental functions, conceptual and otherwise, are dependent on or are very closely related to specific regions, functions, and states of the brain. This being the case, as neurosciences and biomedical technology develop and progress, do you anticipate that it will be possible to modify the brain so that certain types of conceptual processes and mental states do not arise, or that others can be readily introduced or enhanced? Theoretically speaking, of course.

PATRICIA CHURCHLAND: It is very hard to speculate about what the future is going to be like. But I suppose if you really did know in great detail where and how everything was happening in the brain, then I guess you could intervene to change things that you wanted to change either directly or through some indirect intervention. I guess it might be less possible to do so through drugs than to do so directly, in terms of some sort of patterned electrical stimulation. We are talking about changing specific concepts—

DALAI LAMA: Or conceptual processes, not just thoughts.

PATRICIA CHURCHLAND: I see, then I suppose you might also do it by changing the genes and producing a brain that had certain kinds of circuits that we don't now have.

DALAI LAMA: Is it your understanding that there is one discrete part of the brain that is principally responsible for a certain kind of mental state or function? Even if that is the case, that part would presumably not be capable of operating in isolation, but only in interrelationship with the remainder of the brain?

LARRY SQUIRE: I think the neurosciences' consensus now would be that higher brain states, particularly attitudes and ideas, are dependent upon very complex patterned activities in different parts of the brain. And so it would be very difficult to imagine how one could ever direct changes by intervention. Rather, the pattern in representing a certain thought or an idea has a certain stability to it. The brain has such patterns, and the best way to get to that pattern would be from a preceding pattern, which would be another leading thought, attitude, or idea.

ALLAN HOBSON: As a psychotherapist, I think it important here to distinguish the very robust evidence for physical interventions that modify vulnerability to disease—and the clearly established need therefore to develop rational treatments. But projecting into the future, I would agree with Larry Squire's notion that the best way to change ideas is to change ideas. And it doesn't seem likely, in my view, that in the foreseeable future we will have a physical way of isolating a particular ideational set and changing it, chemically for example. Nor do I think that is even desirable.

LEWIS JUDD: If you had such capability, you would not be warranted ethically to use it.

ALLAN HOBSON: Right. The point is that the system is clearly open to interventions of two distinctive kinds. One is a biological intervention, the other is a conceptual intervention. The human brain is really unique in that it has these apparently dual aspects which probably represent higher order complexities of the same structural/functional system.

LEWIS JUDD: No matter how capable we become, no matter with what finesse we might introduce specialized chemical agents, or at what patterned complexity we might introduce electrical stimuli or inject neurotransmitters, or enter certain genetic influences with fibroblast viral vectors and things like that, I doubt that we will ever be able to engineer any kind of intervention relating to a strategy of thinking. It will certainly never get to a point where we can be so specific that we could actually isolate, enhance, or eradicate ideas or concepts.

PATRICIA CHURCHLAND: I don't think we know that we will never get to that point. It was probably only fifty years ago that people said that you will never be able to engineer any genetically new organism.

ALLAN HOBSON: That's true.

LEWIS JUDD: I think it is remote. I don't think intervention will ever be that specific or that geometrically detailed.

DALAI LAMA: Can you anticipate a time in the development of your science that you will be able, depending on your knowledge of an individual's brain, to know whether this person will be hot-tempered or not, be very happy or depressed, or be intelligent or dull, purely in dependence upon your understanding of the brain?

ANTONIO DAMASIO: Theoretically, I think that is perfectly possible. It is not likely to happen in our lifetime, but theoretically, yes.

LEWIS JUDD: It certainly is within the realm of probability.

DALAI LAMA: As you continue to progress in the neurosciences, do you think it is possible that one day by investigating a person's brain you will know what that person is thinking of at the moment?

PATRICIA CHURCHLAND: I think that is going to be very difficult for technical reasons, but not so difficult in principle, in theory.

ANTONIO DAMASIO: Theoretically possible, but rather unfeasible.

DALAI LAMA: But it is possible?

PATRICIA CHURCHLAND: Theoretically possible, but you would have to know not only what was going on in many parts of that person's brain simultaneously and sequentially, but also something about the environment and quite a lot about that person's history.

DALAI LAMA: Up to this time, what percentage of the functioning of the brain do we understand?

ROBERT LIVINGSTON: Half of 1 percent.

ANTONIO DAMASIO: I would say more than that.

LEWIS JUDD: I am not sure. I think we have barely scratched the surface.

PATRICIA CHURCHLAND: If you want to know how the neurons are put together in circuits in order to explain something like perception or being able to move appropriately, then we have got part of the story for those things, but we don't really understand how the brain performs such functions in its characteristically integrative fashion.

LARRY SQUIRE: I think at the level of gross functions, we understand a lot. We can talk, as you have heard, about memory and about vision. But one test of our understanding would certainly be: Can we take any of these computations and put them into a computer so the computer can see or make a decision or talk or turn language into print without a helper. We cannot yet do any of those things. So, obviously, we don't understand in any great detail how the brain is doing such things.

ANTONIO DAMASIO: That may be just not understanding the details. Half of 1 percent is really very little. I think that in terms of mechanisms we understand a bit more. If we can't implement it on a machine, it doesn't mean that we fail in understanding. There is a hardware problem that we obviously cannot possibly solve because it has to do with the evolution of biological systems. So maybe we understand a bit more than is reflected in our capability of implementing our understanding in machines.

LEWIS JUDD: But also, the more we understand, the more it is very clear what we really don't know. You might say that there is an exponential increase in the absence of knowledge.

13. A Buddhist Deconstruction of the Mind's Self

A fundamental tenet of Buddhism is the lack of inherent existence of phenomena, including oneself our normal understanding of things as substantial is an illusion, for they exist truly only as dependently related events. Various dialectic approaches have developed in Tibetan Buddhism, designed to challenge and ultimately to dismantle our reified view of reality. The Dalai Lama several times tries to pursue this type of logical enquiry with the scientists. He seems to be interested in the possibility that a scientific understanding of the brain's processes might lead to a similar deconstruction of illusion.

DALAI LAMA: I would like to know whether, as scientists conducting neuroscientific research in your laboratories, you ever have a spontaneous feeling of affection for the brain itself as you normally feel for living beings as persons? Do you ever have that kind of feeling?

LARRY SQUIRE: Sure.

PATRICIA CHURCHLAND: Yes.

ROBERT LIVINGSTON: Absolutely, yes.

LEWIS JUDD: Highly admire it.

DALAI LAMA: Even though we might *like* certain plants such as these flowers, or this table, or jewelry, or whatever else we may consider precious, that kind of feeling is very different from the feeling of affection or love that you have toward another human being. Underlying that feeling is a very personalized perception of that individual person. Do you ever get that feeling of personal affection when you are looking at the brain?

ANTONIO DAMASIO: That's different.

DALAI LAMA: Is the kind of affection and love that we feel toward another person, as a person, a baseless illusion?

ANTONIO DAMASIO: I would say no.

DALAI LAMA: As a result of scientific investigations, we have more or less refuted the notion of a truly existent self. You can't find a soul, or a human being. You can find the human body, but you can't find the human being *per se,* the person *qua person,* the entity for which we feel compassion and affection. Nor can you find consciousness or awareness apart from the brain. So is our affection and love toward another person as a person baseless? Is that affection, love, and compassion focused upon a nonentity?

ALLAN HOBSON: Brain science is inadequate. Neuroscience is only thirty years old. I believe, however, that we can envisage a time when our knowledge of brains would be elevated to that level. If you imagine the current slope of progress, and the increase in complexity of function which we only recently have come to understand in terms of detailed and comprehensive brain mechanisms, then I can envisage a time at which affection, including human affection, and the entire individual personality of a particular brain, will be revealed.

DALAI LAMA: Would you fall in love with it?

ALLAN HOBSON: You would certainly feel compassion.

LEWIS JUDD: But you would not relate to a brain as you would relate to a person. That is really the question.

DALAI LAMA: This would imply that when a person experiences affection and love for another person, the person who is experiencing affection and love has a philosophical notion that the other individual constitutes a *being* or perhaps has a *soul.* That would be the implication.

If you understood the complexity of the brain to such a degree, is it possible for you to have that response to a brain?

ALLAN HOBSON: He is saying there is something more to the brain than its organic constitution. I agree that such is likely to be discovered through progress in brain science.

In fact, this is not at all what His Holiness intends by this line of argument, as will eventually become clear.

ANTONIO DAMASIO: We need to answer the first question, which was whether or not, working in the laboratory, we feel affection toward the brain. That's the way I interpret your question. Can we feel affection, for instance, for material that we are working with, such as an image of the brain or a slice of the brain? The answer to that is no, at least in my view. What I can feel affection for is a particular individual, a person whom I know.

I should like to tell you a little bit about my experience in that respect. The older I get, the more I feel a great affection and compassion for those individuals who have suffered brain damage and who are subjects in our research. And that feeling is something quite powerful. In fact, in our laboratory, we treat these individuals as our friends. They are very precious human beings, and we grieve over their problems. But when I look at sections of their brains as taken in life by MRI scanning, or actual sections of their brains taken postmortem, I don't feel any affection whatsoever. I am interested in these sections esthetically, in much the same way that I am interested in paintings. And also, particularly, because the brain sections provide very rich information for my work, for my thinking. But no, I don't relate with affectional regard for such artifacts. In fact, to be quite truthful, I relate to them somewhat *less* affectionately than I do to paintings.

DALAI LAMA: In order to save the life of a particular human being, suppose that we undertake an operation on that person's brain or some other part of the body. We may actually cut out some small, damaged portion of the brain. You have no compassion for that part, although it is properly identified as a part of that person's brain. Why is the affection missing? Is it because we took that possibly offending part out, in order to save the life of that person?

Now, you know there exists a person, but if we investigate in a reductionist manner, where and what is that person? What is the human being that you have feelings for? Is that the body? Is that the mind? We may suppose that the brain is the one thing which we consider to be the proper owner of the body. So, such things: brain, mind, soul seem to be essential, whether they all exist or not.

ROBERT LIVINGSTON: I have a suggestion here, Your Holiness. This goes back to our intellectual ancestors, to a contemporary of the Buddha, Hippocrates. Greek physicians in the Hippocratic tradition held that the contract between a doctor and a patient is a "Contract of Philia," or mutual friendship, a contract that includes mutual trust and affection. What this patient/physician contract means is that when a physician encounters someone who requests or otherwise exhibits a need for health advice or care—and this is absolutely without limits of age, sex, nationality, creed, or philosophy—the physician has a professional obligation to do whatever he or she can for that person. If the patient consciously acknowledges the physician and appeals for his or her help, then the bilaterality, the mutuality of the contractual relation of friendship goes automatically into effect. The physician, on the basis of that Contract of Philia, must do everything for that patient that he would do for a friend, and he must not do anything to that patient that he would not do to a friend. In other words, patients are not just diagnostic or therapeutic problems.

LEWIS JUDD: The point is, the physician is dedicating his or her knowledge and skills on behalf of the patient as a totality, as a person, not to some fractional part or organ system. The patient is not just a diseased liver or diseased brain, or whatever. The patient is an integrated, whole person.

DALAI LAMA: That's it. You can amputate the person's arms and legs, yet you know the person is still there, even without arms or legs.

PATRICIA CHURCHLAND: Are you exploring this as an argument for believing that there is a soul?

DALAI LAMA: No, that's not the case. Buddhists don't accept that there is a soul. There is a reductionist approach used in Buddhism in which you do not find the self, that is, you find only *emptiness*. The absence of an intrinsic self *is* emptiness, and emptiness is something to be realized. It is not designed to reintroduce a soul. But we do designate a self. There is a conceptual designation of the self, which relates to the body and to the mind. But it is not something you can localize through reductionist analysis.

ALAN WALLACE: I think the question here that was being asked previously and was not really answered was: If we say that no such thing as a self, or a "soul," exists in and of itself, and moreover, if there is no such thing as a mind existing apart from brain and brain states, can we feel compassion toward a brain and brain states? If we cannot, then is compassion itself a deluded state of mind, inasmuch as you are feeling compassion toward a nonentity?

ANTONIO DAMASIO: In that case, I would say, yes. In that sense, we are generally taking the person as a whole, a brain and a body. We can usefully return to the metaphor I used yesterday, which is that human beings are brains that have a body on their backs. Bodies are intimately and constantly connected to their brains and are represented there.

LEWIS JUDD: They are a person as well.

ANTONIO DAMASIO: The person is a totality.

DALAI LAMA: There is a principle you have introduced and that is the disparity between reality and appearances. Do you think this pertains and that there is an appearance of a person, but in reality no such entity exists?

PATRICIA CHURCHLAND: Yes. At least there is an appearance of a mind, or of a self, but there is no such thing. There is an appearance of a flat earth, but it is no such thing.

LEWIS JUDD: It is an aggregation of functions.

PATRICIA CHURCHLAND: It is also true that His Holiness introduced an interesting paradox. In a way, if we think of the person as a brain, and yet, we can cut out a part of the brain and say, well, that part is not part of the person, and for any part of the brain, we could cut it out and say the same, then how can the person's "self" be the outcome of brain activity?

ALLAN HOBSON: Don't take out too much!

PATRICIA CHURCHLAND: I think that's right. The story is going to have to be, "Don't take out too much." For any given small part, you wouldn't want to identify the "self" with that part. But the brain consists of a distributed collection of integrated parts. And you can disturb it a bit. But, of course, if you take out too much, then you don't any longer have this behaving/interacting/speaking thing for which you can have affection, or which you regard as a person, a self.

DALAI LAMA: According to one system of Buddhist philosophy,[1] it is said that there are two types of phenomena. The first type consists of things that can perform functions, and they are thought to truly exist in their own right. They arise in dependence upon causes and conditions and they produce results. According to this same system, there is another type of phenomena which exists purely as a result of convention or conceptual designations: for example, the borderlines of one country with others. These are simply conventions, pure constructions that exist only because of agreement, and are not capable of performing any function. In this philosophy, a certain category of such phenomena is the mere absence of a specific entity. For example, there is no plant on the palm of my hand. The absence of this plant—

ROBERT LIVINGSTON: There are plants there, though. There are fungi, proper plants, which inhabit all your skin.

PATRICIA CHURCHLAND: There are no daffodils there.

DALAI LAMA: No daffodils. There is an absence here of daffodils. That absence is just a conceptual construct. So the question would be, does that mental construct exist? Maybe this is a metaphysical question that science never talks about.

PATRICIA CHURCHLAND: It is true that there is an absence of daffodils in my hand.

DALAI LAMA: Is that absence of daffodils existent or not?

PATRICIA CHURCHLAND: It is not a thing that exists. An absence of something is not a thing that can interact with something. On the other hand, at least in English, we would say that it is true there is no daffodil there. There could be; in fact, I could imagine a glorious bunch of daffodils there, but the absence of the thing is not a thing.

LEWIS JUDD: But it exists in your mind as you think about it.

PATRICIA CHURCHLAND: I said you could imagine it. But there is a Zeno-style paradox coming up here. It is like saying, "Nobody runs faster than Sarah" and then concluding that there is a *thing*, namely nobody, who does run faster than Sarah.

DALAI LAMA: This relates back to the system of Buddhist philosophy that distinguishes things that perform functions from conceptual constructions, like the absence of daffodils in my hand. So the question is: Does the very absence of a daffodil exist, albeit, not as a thing?

PATRICIA CHURCHLAND: The thought of it can exist, and the thought of it can be identified apart from my brain.

DALAI LAMA: Thought? But is the absence itself a thing?

PATRICIA CHURCHLAND: This is a semantic fuddle, like saying "nothing" is a thing!

LEWIS JUDD: Don't step into the pit!

PATRICIA CHURCHLAND: That's why I keep resisting. It is not a thing but, certainly, its absence exists.

DALAI LAMA: What about certain properties like a dynamic process, the ever-changing nature of a thing? It is not tangible, it is not visible.

PATRICIA CHURCHLAND: How about water heating and getting hotter and hotter? There is a change.

DALAI LAMA: The change of temperature of water. So take water as the entity that has the quality of change, or take fire as the entity that has the quality of change. The property of change is something that exists in the water or in the fire. Take this entity of change—is it findable?

PATRICIA CHURCHLAND: If you mean by change, a thing in itself, no. It is a process. It is something that goes on, so that the fire changes. But we don't want to think of it as . . . this is getting into Zeno territory again. It is really a matter of wordplay.

DALAI LAMA: What is this property of change? Is it a thing, material or otherwise, or is it not? That's the question.

PATRICIA CHURCHLAND: It depends on how you want to think about it.

DALAI LAMA: When we asked about the absence of daffodils on my palm, you said it was not a thing.

PATRICIA CHURCHLAND: True.

DALAI LAMA: There are absences.

PATRICIA CHURCHLAND: There are absences.

DALAI LAMA: It does exist, but it is not there. The property of change, is it a thing as you define thing?

PATRICIA CHURCHLAND: No. There are changes. Things change. But change is not a thing, not stuff, not an object.

LEWIS JUDD: But is it not a property of the object that is exhibiting the process of change?

PATRICIA CHURCHLAND: I think what we are looking for are properties of fire, or properties of fire changing, or properties of elements, or properties of brains. But we are not looking for the property of change itself because there isn't any such object.

DALAI LAMA: For the sake of clarification of terminology, does a "thing" have to be a material phenomenon?

PATRICIA CHURCHLAND: Not by definition, not if, in actual fact, there are nonmaterial phenomena, for example, spiritual phenomena or ghosts. However, spirits and ghosts do not appear really to exist.

DALAI LAMA: In your belief system, are there things that are nonmaterial?

PATRICIA CHURCHLAND: I think the only things that there are are physical things. That is, particles, forces, and so on. But I could be wrong.

ANTONIO DAMASIO: To follow up in relation to the issue of the absence of daffodils, or the issue of the changing temperature, you should not talk about those descriptions as things. But you can nonetheless understand the mechanisms that lead to such experiential events or states. And you can describe them, or account for them, by reduction to a different level of reality. For instance, you can explain why there are changes in the temperature of water by knowing what is happening to the molecules of water. And you can know why, for example, water coming from on high can move a wheel during its fall. So you can explain mechanisms which lead to change. But you cannot talk about them as things. You can talk about them as patterns that can be described.

ALLAN HOBSON: But they are patterns of things.

ANTONIO DAMASIO: They are patterns created by things.

LEWIS JUDD: They are ice crystals.

DALAI LAMA: Did you say patterns are not things? Can you say definitely, within the scientific context, that if something exists it should be findable under analysis? That is, if you seek it out, you should be able to find it?

ANTONIO DAMASIO: If you get to the proper level of analysis.

PATRICIA CHURCHLAND: It may be difficult technically. But if it does exist, then there should be some way, either indirectly, as in the case of electrons, or directly, of being able to get a fix on them. However, one would not *believe* something were real in the absence of empirical evidence.

DALAI LAMA: Just as in science, so within Buddhism.

14. In Conclusion

Building Bridges

ROBERT LIVINGSTON: Your Holiness, I should like to begin with a little story. When I first went to San Francisco, the Golden Gate and Oakland/San Francisco Bay Bridges were being built. People started with the idea of having rapid two-way exchanges across the bay. Specialists studied where to locate the bridges. Some experts, like Willis Bailey at Stanford, forewarned that the Golden Gate Bridge would likely fall down during an earthquake because its foundations would be based on serpentine rocks, unstable under powerful shearing forces. Nevertheless, given public authority and the will to proceed, engineers built huge foundations and anchorages for the suspension systems. They then had tugboats pull two wires from shore to shore for each bridge. When those tenuous threads had been lofted and secured in place, mechanical shuttles propelled themselves back and forth, eventually spinning and sheaving enormous cables from which vertical supports for roadways were suspended.

We are engaged in similar teamwork and expect similarly exciting and progressive developments and mutual advantages from facilitated two-way exchanges. Within our respective traditions, each side has been somewhat chary and perhaps condescending in evaluating the enormous philosophical, psychological, and cultural contributions made by these two disparate conceptions of reality.

You nonetheless have expressed personal interest in bridging the gap between the theoretical and practical knowledge resources of Buddhism and Western life sciences. Fortunately, the idea of bridge-building appeals to resourceful constituents on both sides.

The most significant fact is that both sides were originally founded on the principle that if there is substantial evidence to refute existing assumptions and hypotheses—no matter how long established and widely believed these may be—verifiable evidence will replace them with new explanations. This ensures the progressiveness that distinguishes Buddhist and Western scientific traditions from most ideologies.

At the first Mind and Life Conference, some tenuous threads were strung across. But we have not yet established foundations for deciding what may be the most fruitful questions to ask of one another, what experiments might contribute most effectively to mutual progress, and how we should go about verifying one another's evidence.

DALAI LAMA: Two years ago we had the first Mind and Life dialogue. We have then and now been experimenting with different possible approaches and possible grounds for future dialogues. Sometime in the future it might be quite beneficial if we could concentrate on certain points raised in these two Mind and Life dialogues. At these meetings, points are raised which open new possibilities for pursuing more in-depth research, credible research with precision instruments. With experimental investigation of some of these points, I think we might obtain better evidence and a deeper understanding.

The subjects for such research must be not merely Buddhist scholars but rather practitioners who have some contemplative experience. In the past, some such practitioners have visited laboratories in which their bodily responses were measured. Dr. Benson, for example, performed physiological tests and found that extra energy is mobilized during meditation. Thus meditative experiences gave rise to biological changes within the body. So, I suggest that we proceed like that. I think then we might get a clearer picture, for the scientists as well. Experiment on capable practitioners to analyze their mental states and brain functions—I think maybe that could yield interesting results. Perhaps it could also lead to more questions being raised as well as some being answered. I think that's the path of progress, isn't it?

We have had two dialogues now on Mind and Life that have been fairly diverse in the topics that we have explored. What I would like to do in the future is to pinpoint one topic and really come up with ideas for further research that are more clearly channeled and focused. This is more an objective for group consideration than for me to decide. It is for us to decide among ourselves.

There is another point I should mention. I always believe that each individual human being has some kind of responsibility for humanity as a whole. Particularly, I always believe that as scientists, you have a special responsibility. Besides your own profession, you have a basic motivation to serve humanity, to try to produce better, happier human beings. Whether we understand consciousness or not, we must produce warmhearted persons. That is important. I want to express that. Whenever I meet scientists, I always have to say this.

Through my own profession, I try my best to contribute as much as I can. This proceeds without my being concerned whether another person agrees with my philosophy or not. Some people may be very much against my belief, my philosophy, but I feel alright. So long as I see that a human being suffers or has needs, I shall contribute as much as I can to contribute to their benefit. Scientists and medically qualified people can contribute especially. That's different; that's a particular context. A human being needs to be cared for according to your professional calling. You can contribute; that's your shared professional responsibility.

Especially now in the twentieth century, I think the scientists' role is important and urgent. Scientific evidence and explanations can make something that is unseen into something that becomes manifest and understandable. In contrast, explanations, arguments, and interpretations of mind and life that the religious traditions recommend tend to be based more on subjective experience and advocacy. A scientific explanation can usually be objectified and shared on the basis of instrumental evidence that can detect and measure things otherwise obscure. This makes the scientific explanation more tangible and cognizable than a religious explanation that may be mainly derived from introspective experiences. A scientific explanation, of course, would be more effective.

From within the scientists' circle, it is not clear whether affection and compassion are illusions or real. Often we cannot specifically pinpoint the objects of our compassion, of our projected kindness, the objects of our affection. Anyway, having compassion is something very important throughout human society, isn't it? Whether compassion has an independent existence within the self or not, compassion certainly is, in daily life, I think, the foundation of human hope, the source and assurance of our human future.

Afterword

Buddhist Reflections

B. Alan Wallace

Due to the unusual brevity of this Mind and Life Conference, which lasted two days instead of the five days for all the other meetings in this series, Robert Livingston asked me to write a concluding essay providing further context and elucidation of the Buddhist topics raised here by the Dalai Lama. The following is my attempt to fulfill that wish, principally setting forth certain Buddhist perspectives on the mind/body problem, and at times viewing modern scientific assertions in light of the Buddhist worldview.[1] My motivation in doing so is not to demonstrate the superiority of one view over the other, but to open up new avenues of theoretical and empirical research to scientists and Buddhists alike. For there are, I believe, an increasing number of people today who, like myself feel that modern neuroscience and Buddhism have a great deal to learn from each other. Neither has sole access to exploring the true nature of the mind or body.

The Reality of Suffering

The fundamental structure of Buddhism as a whole is known as the Four Noble Truths. All Buddhist theories and practices are presented within the context of these four, namely the reality of suffering, the reality of the sources of suffering, the reality of the cessation of suffering together with its underlying causes, and finally the reality of the path to such cessation. The Buddha's injunctions regarding these four is that one should recognize the reality of suffering, eliminate the sources of suffering, accomplish the cessation of suffering, and follow the path leading to cessation.

Buddhism identifies two kinds of suffering: physical and mental. The two are not identical, for it is experientially apparent that one may be physically uncomfortable—for instance, while engaging in a strenuous physical workout—while mentally cheerful; conversely, one may be mentally distraught while experiencing physical comfort. This immediately raises the issue of the mind/body relationship. The fact that we have compelling grounds for not simply equating mental and physical degrees of well-being implies a kind of *affective* dualism between the body and mind. Such dualism is explicitly accepted by Buddhism and no reasons were presented in this conference why this should be refuted by modern neuroscience.

Affective dualism may be included in the broader category of what may be deemed *experiential* dualism: our experiences of objective, physical phenomena are quite unlike our experiences of subjective, mental phenomena. An event like an apple dropping from a tree, or a thing like an apple itself, appears quite different from the event of losing hope, or the experience of confidence. Similarly, there are significant experiential differences between objectively observing brain processes and subjectively observing mental processes: the former have specific locations and are composed of material entities that have shape, color, mass, and numerous other physical attributes; mental processes seem to lack those physical attributes, while possessing qualities of their own that are not apparent in brain processes. The fact that Buddhist contemplatives have observed the mind for centuries yet formulated no theory of the brain implies that introspective knowledge of the mind does not necessarily shed any light on the brain. Likewise, the study of the brain alone—independent of all first-person accounts of mental states—does not necessarily yield any knowledge of mental phenomena. Thus, experiential dualism, which maintains that physical and mental phenomena experientially seem to be different, is accepted by Buddhism as well as by at least some of the scientists in this meeting.

Experiential dualism also includes what may be called *causal* dualism, for the mind/body system in Allan Hobson's words "is clearly open to interventions of two distinctive kinds. One is a biological intervention, the other is a conceptual intervention." Lewis Judd concurs when he comments that there "is evidence that there may be a synergistic effect between psychopharmacology and specific forms of psychotherapy." For with the combination of the two, the rate of relief for the clinically depressed is higher than if one administers medications alone. Likewise, Buddhism maintains that the mind is influenced by, and exerts its own influence upon, both mental and physical phenomena.

What shall we make of such mind/body dualisms, which are commonly accepted in Buddhism and in modern science? The Madhyamaka view, which the Dalai Lama endorses and which in Tibet is generally considered the pinnacle of Buddhist philosophy, maintains that humans have an innate tendency to reify both the contents of experience as well as ourselves as experiencing agents. According to this view, while it is useful to recognize the apparent differences between physical and mental events in the above ways, it is a profound error to conclude that nature itself—independently of our conceptual constructs—has created some absolute demarcation between physical and mental phenomena. Thus, the Madhyamaka view explicitly refutes Cartesian substance dualism, which has been so roundly condemned by contemporary neuroscientists. Madhyamikas, or proponents of the Madhyamaka view, declare that if the mind and body did each exist inherently—independently of conceptual designations—they could never interact. Thus, there is a deep incongruity between appearances and reality: while mind and matter seem to be inherently different types of independently existing "stuff," such appearances are misleading; this becomes apparent only by an ontological analysis of the nature of both types of phenomena.[2]

The difficulty of providing any explanation for the causal interaction of the body and mind if the two are regarded as real, separate "things" has been clearly addressed in this conference, and it is a chief reason why the great majority of neuroscientists have adopted a physicalist view of the mind. From a Buddhist perspective, while this step eliminates the need for any causal mechanism relating a nonphysical mind with the brain, it has the disadvantage of shedding no light on the actual nature of consciousness or its origins. Indeed, though modern neuroscience has discovered many elements of the brain and neural processes that are *necessary* for the production of specific conscious processes, it has provided no cogent explanation of the nature of consciousness, nor does this discipline have any scientific means of detecting the presence or absence of consciousness in any organism whatsoever. Over the years since this meeting, I have heard no more illuminating materialist explanation of consciousness than that offered here, namely that it is simply a natural condition of the activated brain. Nor have I heard anything more revealing concerning the origins of consciousness than the statement that it is something that arises when there are enough neurons with elaborate enough connections to support conscious activity. Such accounts actually explain nothing, and they can hardly be counted as scientific theories, for they do not lend themselves to either empirical verification or refutation.

Not only do Madhyamikas reject the notion that the mind is an inherently existent substance, or thing, they similarly deny that physical phenomena as we experience and conceive of them are things in themselves; rather, physical phenomena are said to exist in relation to our perceptions and conceptions. What we perceive is inescapably related to our perceptual modes of observation, and the ways in which we conceive of phenomena are inescapably related to our concepts and languages.

In denying the independent self-existence of all the phenomena that make up the world of our experience, the Madhyamaka view departs from both the substance dualism of Descartes and the substance monism—namely, physicalism—that is characteristic of modern science. The physicalism propounded by many contemporary scientists seems to assert that the real world is composed of physical things-in-themselves, while all mental phenomena are regarded as mere appearances, devoid of any reality in and of themselves. Much is made of this difference between appearances and reality.

The Madhyamaka view also emphasizes the disparity between appearances and reality, but in a radically different way. All the mental and physical phenomena that we experience, it declares, appear as if they existed in and of themselves, utterly independent of our modes of perception and conception. They appear to be *inherently existing things*, but in reality they exist as *dependently related events*. Their dependence is threefold: (1) phenomena arise in dependence upon preceding causal influences, (2) they exist in dependence upon their own parts and/or attributes, and (3) the phenomena that make up the world of our experience are dependent upon our verbal and conceptual designations of them.

This threefold dependence is not intuitively obvious, for it is concealed by the appearance of phenomena as being self-sufficient and independent of conceptual designation. On the basis of these misleading appearances it is quite natural to think of, or conceptually apprehend, phenomena as self-defining things in themselves. This tendency is known as reification, and according to the Madhyamaka view, this is an inborn delusion that provides the basis for a host of mental afflictions. Reification decontextualizes. It views phenomena without regard to the causal nexus in which they arise and without regard to the specific means of observation and conceptualization by which they are known. The Madhyamaka, or Centrist, view is so called for seeking to avoid the two extremes of reifying phenomena on the one hand and denying their existence on the other.

In the Madhyamaka view, mental events are no more or less real than physical events. In terms of our commonsense experience, differences of kind do exist between physical and mental phenomena. While the former commonly have mass, location, velocity, shape, size, and numerous other physical attributes, these are not generally characteristic of mental phenomena. For example, we do not commonly conceive of the feeling of affection for another person as having mass or location. These physical attributes are no more appropriate to other mental events such as sadness, a recalled image from one's childhood, the visual perception of a rose, or consciousness of any sort. Mental phenomena are, therefore, not regarded as being physical, for the simple reason that they lack many of the attributes that are uniquely characteristic of physical phenomena. Thus, Buddhism has never adopted the physicalist principle that regards only physical things as real. To return to the First Noble Truth, both physical and mental suffering are to be recognized, but according the Madhyamaka view, neither exists as a thing-in-itself, and therefore the dualism between them is of a relative, not an absolute, nature.

The Reality of the Sources of Suffering

Just as Buddhism recognizes two types of suffering—mental and physical—so does it affirm the existence of both mental and physical causal influences that give rise to suffering. Physical injury, for example, produces physical pain, and it may also result in mental anguish. On the other hand, certain attitudes such as arrogance, insecurity, craving, hostility, and jealousy may also result in mental distress, and these mental impulses may also lead one into activities that produce physical pain as well. It is also apparent that physical illnesses and injuries do not necessarily result in mental distress—they do not do so for everyone whenever such physical events occur—and mental suffering may arise even in the absence of any apparent physical influences. For example, one may feel deeply distressed by *not* receiving a telephone call from someone. This is not to say that there are no neurophysiological correlates to such unhappiness—that is, that there are no brain events that may be necessary for the arising of unhappiness—but it is not evident that those physical processes are the primary causes of one's distress.

Indeed, Tibetan Buddhism asserts that all the mental states we experience as humans do have physiological correlates in the body, but it does not reduce the subjectively experienced mental states to purely objective, bodily states.

As the Dalai Lama has affirmed many times, if elements of Buddhist doctrine, including the Madhyamaka view, are compellingly refuted by new empirical evidence or cogent reasoning, then those Buddhist tenets must be abandoned. Many neuroscientists today claim that mental processes are in fact nothing other than brain processes: all mental events are either identical to brain events or are solely produced by them and have no existence apart from them. This view is at variance with Buddhism, so if there are compelling grounds for adopting it, Buddhist doctrine should be revised accordingly.

The ever-growing body of neuroscientific discoveries concerning the correspondence of specific mental processes to specific neural events can be reasonably interpreted in either of two ways. This evidence might suggest that mental processes are *identical* to, or at least *concomitant* with, their corresponding brain processes. If this turns our to be the case, this could be regarded as evidence in support of the materialist view that the mind is simply a function of the brain, but this is certainly not the only logical conclusion that could be drawn from such evidence. Alternatively, such correspondences between mental and neural processes might demonstrate that mental processes occur in *dependence* upon brain processes. This suggests a causal relation between two sorts of phenomena, which leaves open the possibility that there may be other causes—possibly of a nonphysical, cognitive nature—that are necessary for the production of mental processes.

Commonsense experience suggests that mental and physical events exert causal influences upon each other. It has long been known that physical stimuli from our environment and from our body influence our perceptions, our thoughts, and feelings. And mental activity—including those same perceptions, thoughts, and feelings—influences the body. Buddhists take such causal interrelatedness at face value; neither physical nor mental causal agency is discounted due to any speculative presuppositions. Buddhism regards subjectively experienced mental events as being nonphysical in the sense that they are not composed of particles of matter; it regards physical events as being nonmental in the sense that they are not of the nature of cognition. Given this limited kind of dualism, what kind of physical mechanism does it posit to account for the causal interaction between these two kinds of phenomena? This question presupposes that all causation requires physical mechanisms, but this assumption has never been held by Buddhists.

It is not evident to me that contemporary physics refutes the limited dualist view proposed by the Madhyamaka view. Modern cosmology suggests that the physical world may have arisen from space itself, which is not composed of particles of matter and hence is not physical in the above Buddhist sense of the term. Many physicists now regard time, too, as being very like a dimension of space, and even energy itself is not necessarily a purely objective, material entity. While the principle of the conservation of energy has often been posited by neuroscientists as a physical law that prohibits any nonmaterial influences in the physical world, Richard Feynman (himself an avowed physicalist) points out that this is a mathematical principle and not a description of a mechanism or anything concrete. He adds that in physics today we have no knowledge of what energy *is*,[3] leaving open the possibility, as the Madhyamikas propose, that energy as we conceive it is not something that exists purely objectively as an independent physical reality. Given the interchangeability of mass and energy, this raises interesting questions concerning the ontological status of matter as well.

The contemporary theoretical physicist Euan Squires explicitly claims that the conservation laws of physics should not be posited as compelling grounds for refuting dualist hypotheses of mind and matter.[4] Until the work of Newton, physicists believed that all forces were simply "push/pull" effects of material bodies, but Newton's law of gravitation countered that the presence of an object at one place could influence the behavior of another at an arbitrarily large distance away, without any intervening medium or mechanism. Thus, as Squires points out, "materialism" in its narrowest interpretation died in the seventeenth century. Similarly, until the late nineteenth century, most scientists viewed the world from the perspective of mechanistic materialism, which required a material medium for the propagation of light. But this principle of mechanism also became obsolete when Maxwell mathematically demonstrated that no such medium was necessary, and Michelson and Morley empirically demonstrated the absence of any physical evidence for such a medium. Thus, the classical principle of mechanism died in the nineteenth century and was even more deeply entombed by twentieth-century discoveries in the field of quantum mechanics.[5] In the examples cited above, speculative preconceptions have been dispelled by advances in knowledge, in the best spirit of scientific inquiry.

To return to the Buddhist account of mind/body interactions, if mental and physical processes do not influence each other by means of some mechanism, how do they interact? Buddhism begins by affirming the validity of our commonsense conclusion that mental and physical phenomena influence each other—a point that the scientists in this conference explicitly confirmed. This affirmation is made on the basis of a very straightforward Buddhist definition of causality: A can be regarded as a cause of B if and only if (1) A precedes B, and (2) were the occurrence of A to have been averted, the occurrence of B would have been averted. Thus, this phenomenological theory of causality does not necessarily require mechanism.

As the Dalai Lama pointed out, there is a simple, twofold classification of causality that has a strong bearing on the nature of consciousness. A may be a *substantial cause* of B, in which case it actually transforms into B, or A may be a *cooperative cause* of B, in which case it contributes to the occurrence of B, but does not transform into it. Now if mental states are in fact nothing other than brain states, then there is no problem in asserting that prior neurophysiological events transform into mental states, and thereby act as their substantial causes. But to conclude with certainty that mental events are *identical* to their neural correlates—or that those mental events are simply a function or state of the corresponding brain states—it would have to be demonstrated empirically that the two occur simultaneously and not sequentially. This would entail knowing the precise moment when a mental event takes place and the precise moment its neural correlate takes place, and ascertaining whether those two moments are simultaneous or sequential. To the best of my knowledge, this has not yet been done, and it is not clear to me how it could be done with sufficient precision. If mental events are produced from prior neural events, the two cannot be identical, in which case it is valid to ask: Do physical processes act as substantial causes or as cooperative causes for mental processes?

If physical events, in causing nonphysical mental events, were to transform into them, the mass/energy of those physical events would have to disappear in the process; this is a position rejected by Buddhism and science alike, albeit for different reasons. Buddhism therefore proposes that physical processes may act as cooperative, but not substantial, causes for mental processes. In the meantime, physical events commonly act as substantial causes for subsequent physical events. But this raises the question: If preceding physical processes act only as cooperative causes for mental events, what, if anything, are the substantial causes of mental events? If mental processes had no substantial causes, this would imply that they arise from nothing; Buddhism rejects this possibility, just as it rejects the notion that physical events can arise from nothing.

The conclusion drawn by Buddhism is that prior mental events act as the substantial causes of subsequent mental events. At times, specific mental states enter a dormant state, as, for example, when visual awareness is withdrawn as one falls asleep. But the continuum of the mind is never annihilated, nor does it ever arise from nothing.

The whole of Buddhism is concerned with identifying the nature and origins of suffering and with exploring means to eliminate suffering from its source. Relying chiefly on contemplative and logical modes of inquiry, it is concerned chiefly with mental afflictions, as opposed to physical illness, and it has attended more to the mental causes of distress than to physical causes. In its pursuit of understanding the physical causes of mental suffering, Buddhism has much to learn from modern neuroscience. There is nothing in Buddhism to refute genetic influences, electrochemical imbalances in the brain, and other types of brain damage as contributing to mental dysfunctions, but in the face of such compelling evidence, a Buddhist might ask such questions as: if two people are genetically prone to a certain type of mental disorder, why is it that one may succumb to the disease and the other not? Likewise, two people may be subjected to very similar kinds of trauma, yet their psychological responses may be very different. To limit the pursuit of such questions solely to physiological causation seems unjustified, regardless of one's metaphysical orientation. The identification of a physical cause of a mental disorder does not preclude the possibility of important psychological factors also being involved. Thus, counseling someone to avoid or more successfully manage certain kinds of circumstances that may lead to mental problems may be sound advice. However, Buddhism is more concerned with identifying and healing the inner mental processes that make one vulnerable to such outer influences. Rather than trying to control or avoid outer circumstances, Buddhism recognizes that many difficult outer circumstances are uncontrollable and at times unavoidable; therefore it focuses primarily on exploring the malleability of the mind, especially in terms of making it less prone to afflictions regardless of one's environment.

In short, Buddhism places a greater emphasis on controlling one's own mind rather than on controlling one's environment. This may be why the Dalai Lama expressed such an interest in the range of causes of such mental disorders as chronic depression, for Buddhism is concerned with counteracting the principal causes of such disorders and not simply with treating their symptoms. For all the medical advances in understanding chronic depression, Lewis Judd candidly acknowledged that antidepressants do not "cure" these disorders; they "treat" or "manage" them as clinicians "try to remove the symptoms." This may be immensely useful in the short term, but for the long term, Buddhism stresses the importance of identifying the necessary and sufficient causes of all kinds of mental disorders with the hope that they may be eliminated and the individual may be utterly healed.

Why is it that medical science so often confines itself to explanations involving physical causation and so swiftly relegates other influences to the euphemistic category of "placebo effects" (bearing in mind that a placebo is defined as something that has no significant medical effects)? I suspect this is largely due to the fact that for the first three hundred years following the Scientific Revolution, there was no science of the mind in the West, and for the first hundred years in the development of psychology, the nature, origins, and causal efficacy of consciousness were widely ignored, with the brief exception of a few introspectionists such as William James. As James commented, those phenomena to which we attend closely become real for us, and those we disregard are reduced to the status of imaginary, illusory appearances, equivalent finally to nothing at all.[6] While the brain has become very real for scientists observing the objective, physical correlates of mental activity, with no comparable development of sophisticated techniques for exploring mental phenomena firsthand, such subjective phenomena as mental imagery, beliefs, emotions, and consciousness itself have been widely regarded as mere illusory epiphenomena of the brain.

Buddhist contemplatives, on the other hand, have long ignored the brain's influence on the mind and therefore attribute little if any significance to it. But they have developed a wide array of introspective, contemplative methods for training the attention, probing firsthand into the nature, origins, and causal efficacy of mental events, including consciousness itself, and for transforming the mind in beneficial ways. Centuries of experience derived from Buddhist practice suggest that the mind may be far more malleable and may hold far greater potential than is now assumed by modern science. However, as the Dalai Lama has commented elsewhere, these claims are like paper money. If we are to attribute value to them, we must be able to verify that they are backed by valid experience. Only that is the gold standard behind the currency of these Buddhist claims.

Does modern cognitive science know enough about the brain and mind to safely conclude that the hypothesis of a nonphysical mind is useless? When asked what percentage of the functioning of the brain we presently understand, neuroscientist Robert Livingston replies, "Half of 1 percent," and Lewis Judd concurs that "we have barely scratched the surface." Nevertheless, one may still hold to a physicalist view of the mind on the grounds that there is no scientific evidence for the existence of any nonphysical phenomena whatsoever, so the hypothesis of a nonphysical mind should not be entertained even for a moment. This would be a very cogent conclusion if science had developed instruments for detecting the presence of nonphysical phenomena and those instruments yielded negative results. However, to the best of my knowledge, no such instruments have ever been developed. Thus, the statement that there is no scientific evidence for the existence of anything nonphysical is unsubstantiated. If neuroscientists had a thorough understanding of all the necessary and sufficient causes for the production of consciousness, and if all those causes turned out to be physical, then all dualist theories of the mind and brain would have to be rejected. But contemporary neuroscientists agree that they are very far from that goal.

It is pertinent to point out here that, strictly speaking, there is still no scientific evidence for the existence of consciousness! Scientists know of its existence only because they are conscious themselves, and they infer on that nonscientific basis that other similar beings are conscious as well. But how *similar* to a human being must another entity be to be deemed conscious? When it comes to the presence or absence of consciousness in unborn human fetuses and in other animals there is no scientific consensus for the simple reason that there is no scientific means of detecting the presence or absence of consciousness in anything whatsoever. This accounts for the current lack of scientific knowledge concerning the nature, origins, and causal efficacy of consciousness. With this in mind, we now turn to the topic of the cessation of suffering and the possibility of the cessation of consciousness itself.

The Reality of the Cessation of Suffering

Once the full range of suffering has been identified and its necessary and sufficient causes discovered, Buddhism asks: Is it possible to be forever freed from suffering and its sources? Many scientists would respond with a swift affirmative: when you die, all your experiences stop, for consciousness vanishes. In other words, the cessation of suffering occurs due to personal annihilation. While this is often promoted as a scientific view, from a Buddhist perspective, the present state of neuroscientific ignorance concerning the origins and nature of consciousness lends little credibility to any conclusions scientists may draw about the effect of biological death on consciousness.

Tibetan Buddhism asserts that during the process of dying, our normal sensory and conceptual faculties become dormant. The end result of this process, when all our normal mental faculties have withdrawn, is not the cessation of consciousness, but rather the manifestation of very subtle consciousness, from which all other mental processes originate. The presence of this subtle consciousness, according to Tibetan Buddhism, is not contingent upon the brain, nor does it entail a loss of consciousness. Rather, the experience of this consciousness is the experience of unmediated, primordial awareness, which is regarded as the fundamental constituent of the natural world. When the connection between this subtle consciousness and the body is severed, death occurs. But this consciousness does not vanish. On the contrary, from it temporarily arises a "mental body," akin to the type of nonphysical body one may assume in a dream. Following a series of dreamlike experiences subsequent to one's death, this mental body also "perishes," and in the next moment one's next life begins, for example in the womb of one's future mother. During the development of the fetus, the various sensory and conceptual faculties are developed in reliance upon the formation of the body. But mental consciousness is said to be present from the moment of conception.

What are the empirical grounds for this theory of metempsychosis, presented here only in outline? Many highly trained Tibetan Buddhist contemplatives claim to be able to recall the events of their previous death, the subsequent dreamlike experience, and the process of taking birth. In many cases they also recall detailed events from their past lives, for the memories are stored, according to this theory, in the continuum of mental consciousness that carries on from one life to another. Other people, too, may have the sense of recalling their past lives, as in the example the Dalai Lama gave of the two girls in India who purportedly recollected the names of people that they had known in previous lives. However, most people do not remember their previous lives, according to Buddhism, for those experiences are eclipsed by the more recent experiences of this life, just as most adults have few memories of their infancy in this life.

In this conference, the scientists' difficulty in understanding the Buddhist concept of subtle consciousness may appear odd, for the notion of subtle physical phenomena is common in science. For example, the electromagnetic field of a single electron is a subtle phenomenon, which can be detected only with very sophisticated instruments. Likewise, the light from galaxies billions of light-years away is very subtle and can be detected only with very powerful, refined telescopes. Similarly, Buddhism posits the existence of subtle states of awareness and mental events that can be detected only with very sensitive, focused, sustained attention. Ordinary consciousness is too unrefined and unstable to detect such phenomena, but Buddhism has devised numerous techniques for training the attention, unknown to modern science, so that it can ascertain increasingly more subtle mental and physical phenomena.[7] While subtle states of awareness can be detected only with very refined awareness, even the grossest mental states, such as rage (which can be ascertained firsthand by an ordinary, untrained mind), cannot be directly detected with the physical instruments of modern neuroscience: they detect only the neurophysiological correlates of such mental states and other related physical behavior. Thus, all states of consciousness may be regarded as too subtle for modern neuroscience to detect.

Whereas belief in an afterlife or the continuity of consciousness after death is often regarded as an optimistic act of faith in the West, Buddhism counters that the belief in the automatic, eternal cessation of suffering at death due to the disappearance of consciousness is an optimistic act of faith, with no compelling empirical or rational grounds to support it. Buddhism does indeed propose that suffering, together with its source, can be radically, irreversibly dispelled, but this requires skillful, sustained refinement of the mind and the elimination of the root cause of suffering—namely, ignorance and delusion—through the cultivation of contemplative insight and knowledge. The means for developing such insight are presented in the Buddhist path to liberation.

The Reality of the Path to the Cessation of Suffering

According to Tibetan Buddhism, the fundamental root of suffering is a type of inborn ignorance regarding the nature of one's own identity, one's own consciousness, and the world of which one is conscious. This tradition claims that all but highly realized people are born with these, but they can be attenuated and even eliminated entirely. Specifically, under the influence of such inborn ignorance we grasp on the absolute duality of self and other, which leads in turn to the reification of all manner of mental and physical phenomena, as well as the division of mental and physical itself. According to the Madhyamaka view, such ignorance is to be countered by realizing the manner in which all phenomena, including oneself, exist as dependently related events as described earlier in this essay.

In addition to such inborn ignorance, human beings are subject to a second type of mental affliction known as speculative ignorance. No one is born with this kind of ignorance, rather it is acquired through false indoctrination and speculation. Buddhism maintains that as a result of adopting unfounded, speculative presuppositions, we may become more confused than we would have been without receiving any formal education whatsoever.

Thus, the proper task of Buddhist training is not to indoctrinate people into a given creed or set of philosophical tenets. Rather, it is to challenge people to examine and reexamine their own most cherished assumptions about the nature of reality. By repeatedly putting our presuppositions to the test of critical examination by way of careful observation and clear reasoning, we empower ourselves to discover and eliminate our own speculative confusion. Once this is cleared away, we are in a much more effective position to detect and vanquish the underlying, inborn ignorance and its resultant mental afflictions. In Buddhism, mental health and spiritual maturation may be measured in direct relation to one's success in overcoming these two types of mental afflictions.

With this twofold classification of ignorance in mind, let us examine the interface between Buddhism and modern science in terms of two quite disparate ways of confronting reality. One is by means of adhering to an ideology and the other is by pursuing scientific inquiry. The eminent anthropologist Clifford Geertz comments in this regard, "Science names the structure of situations in such a way that the attitude contained toward them is one of disinterestedness. . . . But ideology names the structure of situations in such a way that the attitude contained toward them is one of commitment."[8] Geertz regards religious belief as a paradigmatic example of an ideology, and he remarks that this involves a prior acceptance of authority which transforms experience. In short, with respect to any ideology, one who would know must first believe.

The problem of adopting an ideology arises when there is a discrepancy between what is believed and what can be established by compelling evidence. But what constitutes compelling evidence and for whom? Scientists who are committed to physicalism are extremely skeptical of any evidence that is incompatible with that view. As Allan Hobson comments, their minds must be open about such evidence, but that opening is quite narrow. On the other hand, Tibetan Buddhists who are committed to the theory of metempsychosis are extremely skeptical of neuroscientific claims that the mind is simply an epiphenomenon or function of the brain. Thus, with the same neuroscientific evidence presented to them, physicalists find compelling evidence for refuting the nonphysical existence of the mind, whereas traditional Tibetan Buddhists and other nonmaterialists do not.

Most scientists would acknowledge that they do not *know* that consciousness is nothing more than a function of the brain, and most Buddhists, I believe, would acknowledge they do not *know* that consciousness is something more than a function of the brain. And yet convictions run strong in both ways, indicating that both sides are committed to disparate ideologies. If this is true, then scientists, together with Buddhists, may be equally prone to ideologies—or to use Robert Livingston's term, "speculative suppositions." While the history of science is largely an account of disabusing ourselves of mistaken speculative suppositions, as Robert Livingston points out, Buddhism also places a high priority on dispelling such ignorance in order to eliminate the deeper, inborn ignorance that lies at the root of suffering.

Perhaps in order to explore this commonality, the Dalai Lama cited a threefold classification of phenomena that is made in Buddhism. The first of these categories includes phenomena that can be directly apprehended, or empirically demonstrated. The second includes those that are known by logical inference, but not directly. The third includes those that are accepted simply on the basis of someone else's testimony or authority. He hastened to add that these are not qualities inherent to different types of phenomena; rather, they are related to the limitations of our own knowledge. An event that is known to one person solely on the basis of someone else's testimony may be inferentially known by a second person; the same event may be accepted by a third person. Everyone agreed that it is the task of science to reduce the number of phenomena in the third category, and to move as many phenomena as possible from the second to the first category. This, in fact, is the goal of Buddhism as well.

Since it is widely regarded in the West simply as a religion, Buddhist doctrine is still widely regarded as an ideology, in contrast to scientific knowledge. Indeed, many Buddhists do uncritically adopt the tenets of their faith simply as a creed, without subjecting it to either empirical or rational analysis. Ideologies are commonly based not on immediate experience or on cogent, logical analysis, but on the testimony of someone else, such as the Buddha, whom one takes to be an authority. If the words of the Buddha are not accepted as authoritative, then the basis for this ideology vanishes into thin air. Even though many Buddhists do accept Buddhist doctrine in this way, the Buddha admonished his followers: "Monks, just as the wise accept gold after testing it by heating, cutting, and rubbing it, so are my words to be accepted after examining them, but not out of respect [for me]."[9] Thus, unquestioning commitment to an ideology is not only unnecessary in Buddhism, it was explicitly condemned by the Buddha himself!

While scientific knowledge is commonly equated with empirical discoveries, with an ever decreasing reliance upon inference and others' testimony, I believe even a cursory examination of the history of science demonstrates that this view is far from accurate. With the enormous specialization among the sciences and the vast amount of research that has been conducted throughout history and throughout the world today, no single individual can hope to empirically confirm the findings of the rest of the scientific community. Moreover, empirical scientific research relies upon the sophisticated tools of technology, and few scientists have the time or inclination to check the engineering of every instrument they use. For scientific knowledge to progress, scientists must rely *increasingly* on the claims of their scientific and engineering colleagues of the past and present. In most cases, I believe, that trust is well earned, but in most cases that is indeed reliance upon others' authority, not upon one's own observations or rigorous logic. As this is true within the scientific community, it is all the more true for the public at large, which provides the funding for scientific research—people regard scientists as authorities in their respective fields and accept their words on the basis of such trust. This trust is warranted by the belief that *if one were to engage in the necessary scientific training and perform a specific type of research for oneself, one could, in principle, verify other's findings empirically or at least by logical analysis*. It is with this same kind of trust that Buddhist contemplatives receive formal training in Buddhism and try to put to the test the Buddha's own purported discoveries about the nature of suffering, the source of suffering, its cessation, and the path to that cessation.

Buddhist inquiry into the above three types of phenomena proceeds by way of four principles of reason, to which the Dalai Lama referred only briefly in this meeting.[10] To expand briefly on his comments here, the principle of dependence refers to the dependence of compounded phenomena upon their causes, such as the dependence of visual perception upon the optic nerve. It also pertains to the dependence of any type of phenomenon upon its own parts and attributes, or upon other entities, as in the interdependence of "up" and "down" and "parent" and "child." The principle of efficacy pertains to the causal efficacy of specific phenomena, such as the capacity of a kernel of corn to produce a stalk of corn. The principle of valid proof consists of three means by which one establishes the existence of anything: namely, direct perception, cogent inference, and knowledge based upon testimony, which correspond to the above threefold epistemological—and explicitly not ontological—classification of phenomena. The principle of reality refers to the nature of phenomena that is present in their individuating and generic properties. An individuating property of heat, for instance, is heat, and one of its generic properties is that it is impermanent. The Dalai Lama cites as examples of this principle the fact that the body is composed of particles of matter and the fact that consciousness is simply of the nature of luminosity and cognizance. These facts are simply to be accepted at face value: they are not explained by investigating the causes of the body and mind or their individual causal efficacy.

Let us apply these four principles to the materialist understanding of consciousness. According to this view, consciousness is simply a natural condition of the activated brain, much as heat is a natural condition of fire (the principle of reality). As such, consciousness vanishes as soon as the brain is no longer active (the principle of dependence), and it has no causal efficacy of its own apart from the brain (the principle of efficacy). These conclusions are based on the direct observations of neuroscientists investigating mind/brain correlates; they are inferred by philosophers who know of such correlates; and they are accepted as fact by many people who accept scientific materialism without knowing for themselves its supporting empirical facts or logical arguments (the principle of valid proof).

According to the Buddhist view, in contrast, consciousness is simply of the nature of luminosity and cognizance, much as fire is of the nature of heat (the principle of reality). Specific states of consciousness arise in dependence upon the sense organs, sensory objects, and prior, nonphysical states of consciousness (the principle of dependence); and they, in turn, exert influences on subsequent mental and physical states, including indirect influences on the outside physical world (principle of efficacy). These conclusions are purportedly based on the direct observations of contemplatives who have thoroughly fathomed the nature of consciousness; they are inferred by philosophers on the basis of others' experiences; and they are accepted as fact by many Buddhists who accept Buddhist doctrine without knowing for themselves its supporting empirical facts or logical arguments (the principle of valid proof).

In evaluating these two radically different ways of understanding consciousness, the central question arises: which people are deemed to be authorities on consciousness due to their privileged, direct knowledge? Modern Westerners may look with deep skepticism upon anyone claiming to be an authority who is not an accomplished neuroscientist. Traditional Tibetan Buddhists, on the other hand, may look with equal skepticism upon anyone claiming to be an authority on consciousness who has not accomplished advanced degrees of meditative concentration by which to explore the nature of the mind introspectively. By what criteria does one judge who is and who is not an authority who can provide reliable testimony? In other words, whose direct observations are to be deemed trustworthy? I strongly suspect that answers to these questions must address the role of ideology, and perhaps it will turn out to be true that one who would know—either through inference or on the basis of authoritative testimony—must first believe. These questions certainly deserve to be examined in much greater detail, especially in the context of such cross-cultural dialogue.

Before closing, I would like to raise one final issue that is central to Buddhism and to the Dalai Lama himself, and that is compassion. As the Dalai Lama has commented many times, philosophical and religious theories vary from culture to culture, and scientific theories are subject to change over time, but the importance of love and compassion is a constant throughout human history. The Tibetan Buddhist path to liberation and spiritual awakening likewise places an equal emphasis on the cultivation of insight and compassion. Indeed, the experiential knowledge sought in Buddhism is said to support and enhance one's compassion, and any view that undermines compassion is viewed with extreme skepticism.

It was perhaps with this in mind that at one point in this conference the Dalai Lama asked the Western participants whether they—who asserted the identity of the mind (and implicitly the person) with the brain—could feel affection for a brain. The general response among the neuroscientists was perhaps best expressed by Antonio Damasio: "What I can feel affection for is a particular individual, a person whom I know. . . . I don't feel any affection whatsoever [for brains]." Lewis Judd commented in a similar vein, "The physician is dedicating his or her knowledge and skills on behalf of the patient as a totality, as a person, not to some fractional part or organ system. . . . The patient is not just a diseased liver or diseased brain, or whatever. The patient is an integrated, whole person." But where is this "particular individual" or "whole person" to be found? According to physicalism, is this anything more than a baseless illusion, in which case, doesn't this ideology critically undermine love and compassion?

According to the Madhyamaka view, a person cannot be identified with the mind alone or with the brain or the rest of the body. But no individual can be found under analysis apart from the body and mind either. No "I," or self, can be found under such ontological scrutiny, so Madhyamikas conclude, like many neuroscientists today, that the self does not exist objectively or inherently, independently of conceptual designation. However, the Madhyamikas add that while none of us exist as independent things, we do exist in interrelationship with each other. Thus, we do not exist in alienation from other sentient beings and from our surrounding environment; rather, we exist in profound interdependence, and this realization is said to yield a far deeper sense of love and compassion than that which is conjoined with a reified sense of our individual separateness and autonomy.

Whatever fresh insights may be arise from the collaboration of Buddhists and neuroscientists, it is my hope that these may lead us to become more and more "warmhearted persons." I would like to conclude this essay with the Dalai Lama's own concluding words: "Whether compassion has an independent existence within the self or not, compassion certainly is, in daily life, I think, the foundation of human hope, the source and assurance of our human future."

Acknowledgments

Over the years, the Mind and Life Conferences have been supported by the generosity of several individuals and organizations.

Barry and Connie Hershey of the Hershey Family Foundation have been our most loyal and steadfast patrons since 1990. Not only has their generous support guaranteed the continuity of the conferences, but also it has breathed life into the Mind and Life Institute itself.

The institute has also received generous financial support from the Fetzer Institute, The Nathan Cummings Foundation, Mr. Branco Weiss, Adam Engle, Michael Sautman, Mr. and Mrs. R. Thomas Northcote, Ms. Christine Austin, and Mr. Dennis Perlman. On behalf of His Holiness the Dalai Lama and all the other participants over the years, we humbly thank all of these individuals and organizations. Your generosity has had a profound impact on the lives of many people.

We would also like to thank a number of people for their assistance in making the work of the institute itself a success. Many of these people have assisted the institute since its inception. We thank and acknowledge His Holiness the Dalai Lama; Tenzin Geyche Tethong and the other wonderful people of the Private Office of His Holiness; Ngari Rinpoche and Rinchen Khandro, together with the staff of Kashmir Cottage; all the scientists, participants, scientific coordinators, and interpreters; Maazda Travel in the United States and Middle Path Travel in India; Pier Luigi Luisi; Elaine Jackson; Zara Houshmand; Alan Kelly; Peter Jepson; Pat Rockland; Thubten Chodron; Laurel Chiten; Shambhala Publications; Wisdom Publications; and Snow Lion Publications.

The Mind and Life Institute was created in 1990 as a 501(c) 3 public charity to support the Mind and Life dialogues and to promote cross-cultural scientific research and understanding. Visit us at www.mindandlife.org.

Appendix

About the Mind and Life Institute

The Mind and Life dialogues between His Holiness the Dalai Lama and Western scientists were brought to life through a collaboration between R. Adam Engle, a North American businessman, and Dr. Francisco J. Varela, a Chilean-born neuroscientist living and working in Paris. In 1983, both men independently had the initiative to create a series of cross-cultural meetings between His Holiness and Western scientists.

Engle, a Buddhist practitioner since 1974, had become aware of His Holiness's long-standing and keen interest in science, and his desire to both deepen his understanding of Western science and to share his understanding of Eastern contemplative science with Westerners. In 1983 Engle began work on this project, and in the autumn of 1984, Engle and Michael Sautman met with His Holiness's youngest brother, Tendzin Choegyal (Ngari Rinpoche), in Los Angeles and presented their plan to create a week-long cross-cultural scientific meeting. Rinpoche graciously offered to take the matter up with His Holiness. Within days, Rinpoche reported that His Holiness would very much like to participate in such a discussion and authorized plans for the first meeting.

Varela, also a Buddhist practitioner since 1974, had met His Holiness at an international meeting in 1983, the Alpbach Symposia on Consciousness. Their communication was immediate. His Holiness was keenly interested in science but had little opportunity for discussion with brain scientists who had some understanding of Tibetan Buddhism. This encounter led to a series of informal discussions over the next few years; through these conversations,

His Holiness expressed the desire to have more extensive, planned time for mutual discussion and inquiry.

In the spring of 1985, Dr. Joan Halifax, then the director of the Ojai Foundation, and a friend of Varela, became aware that Engle and Sautman were moving forward with their meeting plans. She contacted them on Varela's behalf and suggested that they all work together to organize the first meeting collaboratively. The four gathered at the Ojai Foundation in October of 1985 and agreed to go forward jointly. They decided to focus on the scientific disciplines that address mind and life, since these disciplines might provide the most fruitful interface with the Buddhist tradition. That insight provided the name of the project, and, in time, of the Mind and Life Institute itself.

It took two more years of work and communication with the Private Office of His Holiness before the first meeting was held in Dharamsala in October 1987. During this time, the organizers collaborated closely to find a useful structure for the meeting. Varela, acting as scientific coordinator, was primarily responsible for the scientific content of the meeting, issuing invitations to scientists, and editing of a volume from transcripts of the meeting. Engle, acting as general coordinator, was responsible for fundraising, relations with His Holiness and his office, and all other aspects of the project. This division of responsibility between general and scientific coordinators has been part of the organizational strategy for all subsequent meetings. While Dr. Varela has not been the scientific coordinator of all of the meetings, he has remained a guiding force in the Mind and Life Institute, which was formally incorporated in 1990 with Engle as its chairman.

A word is in order here concerning these conferences' unique character. The bridges that can mutually enrich traditional Buddhist thought and modern life sciences are notoriously difficult to build. Varela had a first taste of these difficulties while helping to establish a science program at Naropa Institute, a liberal arts institution created by Tibetan meditation master Chögyam Trungpa as a meeting ground between Western traditions and contemplative studies. In 1979 the program received a grant from the Sloan Foundation to organize what was probably the very first conference of its kind: "Comparative Approaches to Cognition: Western and Buddhist." Some twenty-five academics from prominent North American institutions convened. Their disciplines included mainstream philosophy, cognitive science (neurosciences, experimental psychology, linguistics, artificial intelligence), and, of course, Buddhist studies. The gathering's difficulties served as a hard lesson on the organizational care and finesse that a successful cross-cultural dialogue requires.

Thus in 1987, wishing to avoid some of the pitfalls encountered during the Naropa experience, several operating principles were adopted that have contributed significantly to the success of the Mind and Life series. These include: choosing open-minded and competent scientists who ideally have some familiarity with Buddhism; creating fully participatory meetings where His Holiness is briefed on general scientific background from a nonpartisan perspective before discussion is opened; employing gifted translators like Dr. Thupten Jinpa, Dr. Alan Wallace, and Dr. Jose Cabezón, who are comfortable with scientific vocabulary in both Tibetan and English; and finally creating a private, protected space where relaxed and spontaneous discussion can proceed away from the Western media's watchful eye.

The first Mind and Life Conference took place in October of 1987 in Dharamsala. The conference focused on the basic groundwork of modern cognitive science, the most natural starting point for a dialogue between the Buddhist tradition and modern science. The curriculum for the first conference introduced broad themes from cognitive science, including scientific method, neurobiology, cognitive psychology, artificial intelligence, brain development, and evolution. In attendance were Jeremy Hayward (physics and philosophy of science), Robert Livingston (neuroscience and medicine), Eleonor Rosch (cognitive science), and Newcomb Greenleaf (computer science). At the concluding session, the Dalai Lama asked that the dialogue continue in biennial conferences. Mind and Life I was published as *Gentle Bridges: Conversations with the Dalai Lama on the Sciences of Mind*, edited by Jeremy Hayward and Francisco Varela (Boston: Shambhala Publications, 1992). The volume has been translated into French, Spanish, Portuguese, German, Japanese, Chinese, and Thai.

Following Mind and Life II (the subject of the present volume), Mind and Life III was held in Dharamsala in 1990. Daniel Goleman (psychology) served as the scientific coordinator. He chose to focus on the relationship between emotions and health. Participants included Dan Brown (experimental psychology), Jon Kabat-Zinn (medicine), Clifford Saron (neuroscience), Lee Yearly (philosophy), and Francisco Varela (immunology and neuroscience). Mind and Life III was published as *Healing Emotions: Conversations with the Dalai Lama on Mindfulness, Emotions, and Health*, edited by Daniel Goleman (Boston: Shambhala Publications, 1997). That volume has been translated into French, Spanish, Portuguese, German, Japanese, Chinese, Dutch, Italian, and Polish.

During Mind and Life III a new mode of exploration emerged: participants initiated a research project to investigate the neurobiological effects of meditation on long-term meditators. To facilitate such research, the Mind and Life network was created to connect other scientists interested in both Eastern contemplative experience and Western science. With seed money from the Hershey Family Foundation, the Mind and Life Institute was born. The Fetzer Institute funded two years of network expenses and the initial stages of the research project. Research continues on various topics such as attention and emotional response.

The fourth Mind and Life Conference met in Dharamsala in October 1992, with Francisco Varela again acting as scientific coordinator. The dialogue focused on the areas of sleep, dreams, and the process of dying. Participants were Charles Taylor (philosophy), Jerome Engel (medicine), Joan Halifax (anthropology, death and dying), Jayne Gackenbach (psychology of lucid dreaming), and Joyce McDougall (psychoanalysis). The account of this conference is now available as *Sleeping, Dreaming and Dying: An Exploration of Consciousness with the Dalai Lama*, edited by Francisco J. Varela (Boston: Wisdom Publications, 1997). That volume has been translated into French, Spanish, German, Japanese, and Chinese.

Mind and Life V was held again in Dharamsala in October 1994. The topic was altruism, ethics, and compassion, with Richard Davidson as the scientific coordinator. In addition to Dr. Davidson, participants included Nancy Eisenberg (child development), Robert Frank (altruism in economics), Anne Harrington (history of science), Elliott Sober (philosophy), and Ervin Staub (psychology and group behavior). The volume covering this meeting, *Visions of Compassion*, was published by the Oxford University Press in December 2001.

Mind and Life VI opened a new area of exploration beyond the previous focus on life sciences. The meeting took place in Dharamsala in October, with Arthur Zajonc as the scientific coordinator. This conference focused on the new physics and cosmology. The participants, in addition to Dr. Zajonc, were David Finkelstein (physics), George Greenstein (astronomy), Piet Hut (astrophysics), Tu Weiming (philosophy), and Anton Zeilinger (quantum physics). The volume covering this meeting, *The New Physics and Cosmology*, was published by Oxford University Press in March 2004.

The dialogue on quantum physics was continued at a smaller meeting held at Anton Zeilinger's laboratories at the Institute for Experimental Physics in Innsbruck, Austria, in June 1998. Present were His Holiness, Drs. Zeilinger and Zajonc, and interpreters Wallace and Jinpa. That meeting was written up in the cover story of January 1999 issue of GEO magazine of Germany.

In March 2000, the next meeting was held in Dharamsala, with Daniel Goleman as scientific coordinator. The discussion returned to cognitive sciences, with a focus on destructive emotions.

Notes

Introduction

1. Published as *Gentle Bridges: Conversations with the Dalai Lama on the Sciences of the Mind*, ed. Jeremy Hayward, PhD and Francisco Varela, PhD (Boston and London: Shambhala Publications, 1992).
2. The third conference in 1991 was published as *Healing Emotions: Conversations with the Dalai Lama on Mindfulness, Emotion, and Health*, ed. Daniel Goleman, PhD (Boston and London: Shambhala Publications, 1997); the fourth conference in 1993 was reported in *Sleeping, Dreaming, and Dying: An Exploration of Consciousness with the Dalai Lama*, edited and narrated by Francisco J. Varela, PhD (Boston: Wisdom Publications, 1997).

Chapter 2: Toward a Natural Science of the Mind

1. Robert Livingston elaborates: "According to Descartes, the nervous system was considered rather as an automaton. He refers to the pineal gland in the center of the brain as being in fact the seat of the rational soul in humans. It received information out of which it could understand the world. And out of the pineal gland arose the voluntary, 'rational,' control of the body commanding it to behave accordingly. Other aspects of behavior were automatic. He made a clear distinction with animals who shared an animating, but not a rational, soul."
2. According to Buddhist psychology, memories are stored in the mind. The mental processes that are dependent on the brain for bringing those memories to consciousness are impaired because of brain dysfunction, but those memories may be recalled in a future life.

Chapter 3: A Buddhist Response

1. Werner Heisenberg, *Physics and Philosophy: The Revolution in Modern Science* (New York: Harper and Row, 1962), p. 58.

Chapter 4: The Spectrum of Consciousness: From Gross to Subtle

1. The source of these four types of analysis is the *Saṃdhinirmocana Mahāyāna Sūtra*, translated into English as *Wisdom of the Buddha* by John Powers (Berkeley: Dharma Publishing, 1995), pp. 285–95, 367–69. The analysis of dependence concerns the dependence of compounded phenomena in terms of their arising due to their prior causes and conditions. The analysis of the performance of functions concerns the causes and conditions for the ascertainment of phenomena, their establishment as being existent, or their activities. The analysis of logical correctness concerns the causes and conditions for individual understanding, explanation, establishing the meaning of propositions, and comprehension. Such analysis is pursued on the basis of immediate perception, logical inference, and knowledge based on reliable authority. The analysis of reality refers to the nature of phenomena that is present in specific and general characteristics. Examples are the facts that fire burns, water moistens, and wind moves.

Chapter 7: Steps toward an Anatomy of Memory

1. Located in the parahippocampal gyrus, the perirhinal cortex, and the entorhinal cortex.

Chapter 8: Brain Control of Sleeping and Dreaming States

1. An account of this conference is presented in *Sleeping, Dreaming, and Dying: An Exploration of Consciousness with the Dalai Lama*, edited and narrated by Francisco J. Varela, PhD (Boston: Wisdom Publications, 1997).
2. Dr. Hobson is referring to the dream yoga teachings that instruct the practitioner to sleep on the right side.

Chapter 11: Psychiatric Illnesses and Psychopharmacology

1. *The Diagnostic and Statistic Manual of Mental Disorders*, American Psychiatric Association (Washington DC: American Psychiatric Press, 1994).

Chapter 13: A Buddhist Deconstruction of the Mind's Self

1. Specifically, the Sautrāntika system, which advocates a kind of metaphysical realism and substantial dualism that is refuted in the Madhyamaka school, which is widely regarded as the pinnacle of Buddhist thought in Tibet.

Afterword: Buddhist Reflections

1. I would like to thank Zara Houshmand, Jose Cabezón, and Geshe Thupten Jinpa for their helpful comments on this essay.
2. This is not to deny that on a relative, or conventional level, Madhyamikas do assert the duality of mental and physical phenomena. However, according the Buddhist tantric view adopted by the Dalai Lama and most other Tibetan Buddhists, this duality is not as radical as the duality, say, of sentient and nonsentient entities. The reason for this is that all mental events fundamentally stem from the "very subtle energy-mind" (Tib. *shin tu phra ba'i rlung sems*), which is a primordial reality having both physical and cognitive attributes. The physical world, too, is said to be a creative display (Tib. *rtsal* or *rol pa*) of this same energy-mind. Thus, according to the Vajrayana, the conventional dualism of mind and matter is based upon a fundamental monism.
3. R. P. Feynman, R. B. Leighton, and M. Sands, *The Feynman Lectures on Physics* (Reading, Mass.: Addison-Wesley, 1963), p. 42.
4. Euan Squires, *Conscious Mind in the Physical World* (Bristol: Adam Hilger, 1990), p. 22.
5. I have delved into this topic from a Buddhist perspective in *Choosing Reality: A Buddhist View of Physics and the Mind* (Ithaca: Snow Lion, 1996).
6. William James, "The Perception of Reality" in *The Principles of Psychology* (New York: Dover Publications, 1950), pp. 290–91.
7. These techniques are discussed in detail in *The Bridge of Quiescence: Experiencing Tibetan Buddhist Meditation* by B. Alan Wallace (Chicago: Open Court, 1997).
8. Clifford Geertz, *The Interpretation of Cultures* (New York: Basic Books, 1973), pp. 230–31.
9. This verse, often quoted in Tibetan Buddhist literature, is cited from the *Vimalaprabhā* commentary on the *Kālacakra*, although it appears

in the Pāli Canon as well. The Sanskrit occurs as a quotation in *Tattvasaṃgraha*, ed. D. Shastri (Varanasi: Bauddhabharati, 1968), k. 3587.

10. The Dalai Lama addresses these four principles at somewhat greater length in his work *The World of Tibetan Buddhism* (Boston: Wisdom, 1995), pp. 47–49; a yet more detailed discussion is found in Matthew Kapstein's essay "Mi-pham's Theory of Interpretation" in *Buddhist Hermeneutics*, ed. Donald S. Lopez Jr. (Honolulu: University of Hawaii Press, 1988), pp. 152–61.

Books by the Dalai Lama

Core Teachings of the Dalai Lama series

The Complete Foundation: The Systematic Approach to Training the Mind
An Introduction to Buddhism
Our Human Potential: The Unassailable Path of Love, Compassion, and Meditation (*forthcoming*)
Perfecting Patience: Buddhist Techniques to Overcome Anger
Stages of Meditation: The Buddhist Classic on Training the Mind (*forthcoming*)
Where Buddhism Meets Neuroscience: Conversations with the Dalai Lama on the Spiritual and Scientific Views of Our Minds

(*more forthcoming*)

Also available from Shambhala Publications

Answers: Discussions with Western Buddhists
The Bodhisattva Guide: A Commentary on *The Way of the Bodhisattva*
The Buddhism of Tibet
Dzogchen: Heart Essence of the Great Perfection
From Here to Enlightenment: An Introduction to Tsong-kha-pa's Classic Text *The Great Treatise on the Stages of the Path to Enlightenment*
The Gelug/Kagyu Tradition of Mahamudra
The Great Exposition of Secret Mantra, Volume 1: Tantra in Tibet
The Great Exposition of Secret Mantra, Volume 2: Deity Yoga
The Great Exposition of Secret Mantra, Volume 3: Yoga Tantra
The Heart of Meditation: Discovering Innermost Awareness
Kindness, Clarity, and Insight
The Pocket Dalai Lama
The Union of Bliss and Emptiness: Teachings on the Practice of Guru Yoga

Index